Heavy-Ion Collisions and Multiparticle Production

Heavy-Ion Collisions and Multiparticle Production

Guest Editors

Qun Wang
Zuotang Liang
Enke Wang

MDPI

Basel • Beijing • Wuhan • Barcelona • Belgrade • Novi Sad • Cluj • Manchester

Guest Editors
Qun Wang
Department of Modern Physics
University of Science and Technology of China
Hefei
China

Zuotang Liang
Institute of Frontier and Interdisciplinary Science
Shandong University
Qingdao
China

Enke Wang
Institute of Quantum Matter
South China Normal University
Guangzhou
China

Editorial Office
MDPI AG
Grosspeteranlage 5
4052 Basel, Switzerland

This is a reprint of the Special Issue, published open access by the journal *Symmetry* (ISSN 2073-8994), freely accessible at: https://www.mdpi.com/journal/symmetry/special_issues/heavy_Ion_collision_multiparticle_generation.

For citation purposes, cite each article independently as indicated on the article page online and as indicated below:

Lastname, A.A.; Lastname, B.B. Article Title. *Journal Name* **Year**, *Volume Number*, Page Range.

ISBN 978-3-7258-4215-5 (Hbk)
ISBN 978-3-7258-4216-2 (PDF)
https://doi.org/10.3390/books978-3-7258-4216-2

Cover image courtesy of STAR Collaboration

Contents

About the Editors

Qun Wang

Qun Wang obtained his Bachelor of Science in physics at Shandong University in 1988, and Ph.D in particle and nuclear physics at Shandong University in 1997. From July 1997 to January 2000, he was a Lecturer and then an Associate Professor in the Department of Physics, Shandong University. From February 2000 to May 2003, he was a Humboldt Fellow in the Institute for Theoretical Physics, University of Frankfurt, Germany. From June 2003 and June 2005, he was a Visiting Professor in the GSI Virtual Institute, a joint research program between the Institute of Theoretical Physics of University of Frankfurt and the German National Laboratory for Heavy Ion Research. He was a senior visiting professor at Brookhaven National Lab in 2014 and at McGill University in 2024. From 2005 to 2021, he was a Professor in the Department of Modern Physics, University of Science and Technology of China (USTC). He has been a Chair Professor in physics at USTC since 2022.

Dr. Wang was selected to join the 100 Talent program of Chinese Academy of Sciences in 2007 and was awarded the Distinguished Young Scholars in 2012 by the National Science Foundation of China (NSFC). He was awarded the Bao-Gang Excellent Teacher's Prize in 2022.

Dr. Wang's research field is theoretical high-energy nuclear physics. His research interest is in the application of quantum field theory in dense and hot matter, phenomenology of heavy-ion collision, and non-equilibrium dynamics of strong interaction matter.

Zuotang Liang

Zuo-tang Liang obtained both his Bachelor of Science in Physics (1984) and Master of Science in physics (1987) from Shandong University under the supervision of Professor Qu-bing Xie. He obtained his Ph.D in 1994 from Free University Berlin under the supervision of Professor Ta-chung Meng. In 1997, he took the position of Vice Professor and then became Full Professor at Shandong University. Currently, he is a Chair Professor at Shandong University.

Zuo-tang's research field is particle and nuclear physics theory. He is particularly interested in QCD high-energy spin physics such as the spin structure of nucleon, spin effects in fragmentation, and spin polarizations in high-energy reactions including heavy-ion collisions. He was awarded the National Research Fund for Excellent Young Scientists in 2005 and was appointed Chang-Jiang Chair Professor in 2009.

Enke Wang

Enke Wang received his Ph.D in theoretical physics from Central China Normal University (CCNU) in 1990. From September 1990 to July 1997, he worked at Jingzhou Teacher's College as Lecturer and Professor. Between July 1997 and December 1998, he served as a Humboldt Research Fellow at the University of Regensburg in Germany. From December 1998, he held the position of Full Professor at the Institute of Particle Physics at CCNU. In March 2002, he was promoted to Dean at the College of Physical Science and Technology at CCNU, and in October 2010, he took on the position of Director of the Graduate School at the same university. From July 2012 to September 2017, he served as Vice President of CCNU, and from September 2017 to December 2023, he was President at South China Normal University.

Prof. Wang has been recognized with several prestigious awards, including the Cheung Kong Scholar Distinguished Professorship in 2008 and the Distinguished Young Scholar award from the National Science Foundation of China (NSFC) in 2009. He is also a two-time recipient of the First

Prize of Natural Science from the Ministry of Education in China, awarded in 2005 and 2019.

Prof. Wang's research focuses on theoretical high-energy nuclear physics, with particular emphasis on finite-temperature field theory and hard probes of quark–gluon plasma.

Preface

In recent decades, the field of high-energy nuclear physics has witnessed remarkable progress in our understanding of the properties of strong interaction matter under extreme conditions. Heavy-ion collisions in particular serve as a tool for exploring quark–gluon plasma, a new state of matter, and the dynamics of multiparticle production. This Special Issue of *Symmetry*, entitled "Heavy-Ion Collisions and Multiparticle Production," is dedicated to the memory of four Chinese pioneers in this field, namely Hong-Fang Chen, Lian-Shou Liu, Ru-Keng Su, and Qu-Bing Xie, who made early and significant contributions to the development of high-energy nuclear physics in China.

Prof. Hong-Fang Chen (1938–2017) was an esteemed educator and experimental physicist who played an essential role in international experiments such as the BES, STAR, and CMS through her leadership and expertise in the technology of high-energy particle detectors. Her dedication to education and mentorship has inspired many students and researchers, making her a revered figure in the field of experimental high-energy physics. She is one of the founders of the first high-energy physics laboratory at USTC, and has made USTC one of the most important training bases for talented young people in particle and nuclear physics in China.

Professor Lian-Shou Liu (1932–2009) was a pioneer in the theoretical study of multiparticle dynamics in high-energy heavy-ion collisions in China. His work on multiparticle dynamics and fluctuations in high-energy nucleus–nucleus collisions provided insights into particle production mechanisms. As the founder of the Institute of Particle Physics at Central China Normal University, he played a central role in fostering international collaborations and establishing China as a key player in high-energy nuclear physics research.

Professor Ru-Keng Su (1938–2022) was a theoretical physicist, with extensive contributions to nuclear and particle physics, black hole physics, and cosmology. His research on gauge fields, quark confinement, and the thermodynamics of nuclear and quark matter has significantly advanced our understanding of strong interaction matter. His work on the quasiparticle model and finite-temperature field theory has been particularly influential. As an educator, he shaped the careers of numerous students who have graduated from Fudan University through his engaging teaching style and dedication to academic excellence.

Professor Qu-Bing Xie (1935–2013) was a theoretical physicist working on multiparticle production and heavy-ion collisions. His work on the quark production rule and the Shandong model has deepened our understanding of hadronization and particle production mechanism in high-energy collisions. He has been instrumental in building up the theoretical particle physics research group at Shandong University, which is now an integral part of the Key Laboratory of Particle Physics and Particle Irradiation (Ministry of Education). Many of his students have gone on to become active researchers in the field.

This Special Issue aims to honor their contributions to high-energy nuclear physics in China as scientists and educators. This collection of papers explores recent advances in high-energy nuclear collisions, quark–gluon plasma properties, and multiparticle production mechanisms. We hope that this collection will not only pay tribute to the achievements of these pioneers, but also inspire future generations of scientists to continue their efforts to understand the nature of the strong interaction.

We are grateful to the authors who have contributed to this Special Issue and to the reviewers who have ensured the high quality of the papers published.

Yu-Gang Ma
Professor of Physcis
Academician of CAS
Institute of Modern Physics
Fudan University
Shanghai, China

Qun Wang, Zuotang Liang, and Enke Wang
Guest Editors

symmetry

MDPI

Article

Quark Matter at High Baryon Density, Conformality and Quarkyonic Matter

Larry McLerran

Institute for Nuclear Theory, University of Washington, Box 351550, Seattle, WA 98195, USA; mclerran@mac.com

Abstract: This paper discusses high-baryon-density quarkyonic matter in the context of recent observations concerning neutron stars and the qualitative reasons why quarkyonic matter explains certain features of the equation of state that arises from these observations. The paper then provides a qualitative discussion of the quarkyonic hypotheses, and the essential features of quarkyonic matter that explain the outstanding features of the equation of state.

Keywords: quarkyonic matter; high baryon density QCD; conformality

1. Introduction

The physics of high-energy-density matter and heavy-ion collisions has been thoroughly developed over the past 40 years. The recent developments associated with the quarkyonic hypothesis will be discussed, concerning the properties of matter at high baryon density and low temperature. Such matter may occur in neutron stars and intermediate-energy nuclear collisions. There is a wealth of exciting new information concerning neutron stars coming from gravitational wave experiments and neutron star radii measurements, providing strong constraints on the properties of such matter. Furthermore, there is the possibility of extracting complementary information from existing facilities by colliding intermediate-energy heavy ions.

The purpose of this paper is to discuss the motivations of the quarkyonic hypothesis based on the theory of strong interactions, QCD, and the properties of recent neutron star observations. First, recent neutron star observations are discussed, in particular, the extracted equation of state and its conformal properties. Next, the quarkyonic hypothesis is proposed and argued that it has the properties needed to describe the qualitative features of the equation of state. Then, the recently conjectured possible experimental probes in low-energy heavy-ion experiments are discussed. The paper ends with a brief outline of some important theoretical issues which are currently poorly understood concerning quarkyonic matter.

The purpose of this paper is to not only provide a comprehensive review of the properties of high-density, low-temperature baryonic matter, but to also provide a simple and qualitative description of quarkyonic matter and its implications in light of recent observations.

2. The Importance of Neutron Star Studies in Nuclear Physics

The measurements of neutron stars can provide their radii and masses, and in the case of gravitational wave detection, constraints on the quadrupole deformability. These properties can be determined by knowing an equation of state,

$$P = P(\epsilon) \quad (1)$$

where the energy density of cold matter is ϵ and the pressure is P. The equation of state plus the Tolman–Oppenheimer–Volkov equation and the general relativistic equation of hydrostatic equilibrium, can determine these quantities. The equation of state needs to be known up to an energy density one order of magnitude greater than that of nuclear matter.

For energy densities twice the nuclear matter energy density, the equation of state is determined by low-energy nuclear theory, and its generalizations involving effective field theory. At two orders of magnitude greater than the nuclear matter energy density, the equation of state is that of quarks, and is determined by QCD computations [1,2]. At intermediate-energy densities, the equation of state is known from first principle computations. Nevertheless, combining the neutron star observations together with the known properties at high- and low-energy density, it is possible to obtain a good determination of the equation of state.

This empirically determined equation of state is important [3–7]. At nuclear matter density, matter is a non-relativistic liquid of nucleons with a sound velocity $v_s^2 \sim 10^{-2} - 10^{-1}$. To describe neutron stars by a density four times that of nuclear matter, the sound velocity is $v_s^2 \geq 1/3$, where $v_s^2 = 1/3$ is the limit for a relativistic ideal gas. There are indications that the sound velocity exceeds 1/3 in this region. This transition from a non-relativistic fluid to an ultra-relativistic system occurs over a very small range of change of typical particle separation, $r/r_0 \sim (\rho/\rho_0)^{-1/3} \sim 1.6$, such that the transition occurs incredibly rapidly.

The computation of the sound velocity as a function of density [8] is shown in Figure 1. This computation involves many fits to neutron star data, and the probability distribution for various sound velocities is shown (Figure 1).

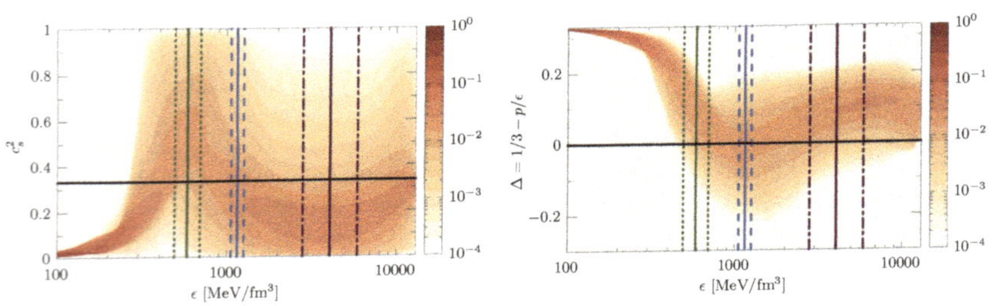

Figure 1. Probability density functions (PDFs) of the speed of sound (**left panel**) and trace anomalies (**right panel**) as functions of the energy density. Vertical lines represent the median and 1σ credibility region for the peak position in c_s^2 (green solid and dotted lines), values at the centre of maximally massive neutron stars (blue solid and dashed lines), and the peak position in χ_B (purple solid and dash-dotted lines). Horizontal, black lines show the conformal values of $c_s^2 = 1/3$ (**left panel**) and $\Delta = 0$ (**right panel**) [8].

Another dimensionless parameter that characterizes the equation of state is the trace omaly scaled by the energy density [9],

$$\Delta = \frac{1}{3} - \frac{P}{\epsilon}. \tag{2}$$

e anomaly is a measure of the deviation from the scale invariance. Determination led trace anomaly is shown in Figure 1. It is notable that the approximate scale can be achieved at densities found inside neutron stars, particularly because ities it is commonly believed that nuclear matter still strongly interacts. Of matter must asymptotically become scale invariant because of the asymptotic ty of QCD; however, this observed precocious scaling of dense matter is . In Figure 2, a contour plot extracted from the analysis of fits to neutron tate in neutron star cores is shown. It is notable that the values centre ariant $v_s^2 = 1/3$ and $\Delta = 0$.

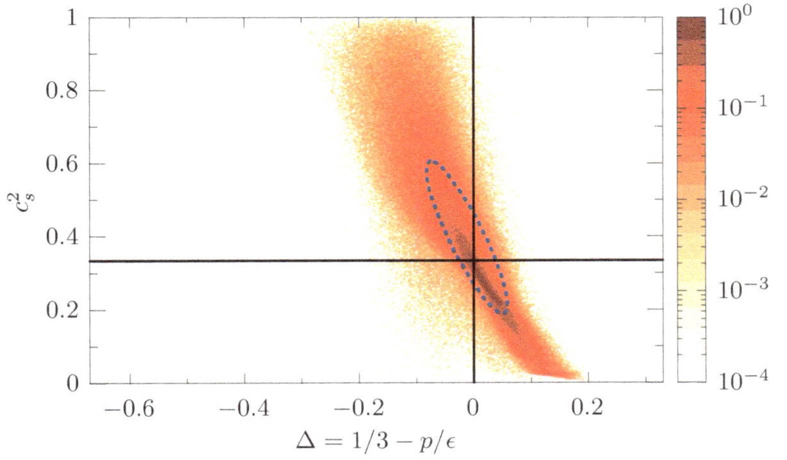

Figure 2. Probability density function (PDF) of the trace anomaly vs. the speed of sound at the centre of maximally stable neutron stars. The dotted blue ellipse marks the 1σ credibility around the mean. The black horizontal and vertical lines mark the conformal values [8].

The trace anomaly and sound velocity are related by

$$v_s^2 = \frac{1}{3} - \Delta - \epsilon \frac{d\Delta}{d\epsilon}. \quad (3)$$

The fact that the sound velocity is large at relatively low densities is therefore another indication that the matter is extremely rapidly approaching the scale invariant limit $\Delta = 0$.

3. Strongly Interacting Matter and the Quarkyonic Hypothesis

The indications described above suggest that matter rapidly approaches an almost conformally invariant distribution of quarks and gluons at densities of about one order of magnitude greater than the density of nuclear matter. Suppose that matter converts between a system of nucleons and quarks at some fixed baryon density. Here the case of isospin zero matter is considered for simplicity, and made from neutrons and protons only (or up and down quarks). To qualitatively understand this transition, let us consider free gasses. The Fermi momenta are denoted as k_F^N and k_F^Q for nucleons and quarks, respectively. We will take the number of quark colours as a parameter to illustrate this point. Since a nucleon is composed of N_c quarks, we might naively expect that $k_F^N = N_c k_F^Q$. On the other hand, the baryon number density is

$$n_B \sim (k_F^N)^3 \sim (k_F^Q)^3. \quad (4)$$

If we take the Fermi momenta of nucleonic matter at a few times the nuclear matter density to be of order Λ_{QCD}, then in the quark matter phase, the Fermi momenta of quarks is also of order Λ_{QCD} and if we try to imagine nucleons made of quarks in this new phase, then they would be relativistic. How can matter make such a drastic change?

A way to resolve such a paradoxical situation is through quarkyonic matter [10,11]. In QCD, nucleons are made of quarks, and discussing which degrees of freedom are quarks and nucleons reflect the desired approximations. At low baryon density, it is a good approximation to treat nucleons as a Fermi gas of nucleons as they occupy low-momentum states. These nucleons are composed of quarks, but the quarks are spread over a momentum scale, which at low momentum is of the order of the QCD scale. However, as one adds more nucleons, these quark states approach an occupation number of order one and this fills the quark states up to a Fermi energy of order of the QCD scale. New nucleons need to be added, but to avoid Pauli blocking of the quarks, they must be added at the Fermi surface corresponding to a nucleon energy of order $N_c \Lambda_{QCD}$. The nucleons and quarks at this point are relativistic. Therefore, there is a transition at a density of order Λ^3_{QCD}, where the degrees of freedom change rapidly from non-relativistic to relativistic.

One can understand quarkyonic matter at high densities, as a Fermi sea of quarks is approximately filled with a Fermi surface of nucleons. However, quark–hadron duality requires that we look at this situation a little closer [12]. A filled Fermi sea of quarks will comprise one baryon for each N_c quarks. It takes N_c quarks to make one baryon. However, at a nucleon momentum of order $N_c k_{quark}$, the occupation number of a filled Fermi gas is N_c^3 times the occupation number of a quark because the De Broglie wavelength of a nucleon is $1/N_c$ compared to a quark. Therefore, the nucleon occupation number with $k_N = N_c k_Q$ composed of quarks in the Fermi sea is $1/N_c^3$. This is true until one approaches the Fermi surface for quarks. Here, one can completely fill nucleon states close to the Fermi energy, $k_F^N = N_c k_F^Q$. The quark constituents are spread out in momentum above the Fermi sea because of the intrinsic momentum of quarks inside the nucleon, making a Fermi surface for quarks whose width is $\Delta k \sim \Lambda_{QCD}$. Therefore, the picture of quarks that arises is a filled Fermi sea of quarks with a Fermi surface of width Λ_{QCD}. This can also be thought of in terms of nucleons as an under-occupied distribution of nucleons with an occupation number of order $1/N_c^3$ followed by this surface of fully occupied nucleons. The nucleon distribution is linked with the quark distribution. This is the under-occupied distribution of nucleons that corresponds to a filled Fermi sea.

Let us summarize this argument in a slightly different way. When we think about quarkyonic matter we must have a dual picture in mind. One picture is the distribution of quarks and the other is the distribution of nucleons. These pictures must be consistent with one another. The quark distribution is a filled Fermi sea of quarks plus a tail above some momentum that falls to zero. In the tail, the occupation number of quarks falls to zero, and its width is of order of the scale of quark momentum in the nucleon wavefunction. In this tail, we can fit nucleon states up to some saturation density. The total density of nucleons is of the same order as when the quark sea first appears, but the local phase space density of nucleons, dn_N/d^3k, is saturated and of order 1. Thus, the nucleons form a shell with a typical momentum $k_N^F \sim N_C k_q^N$. On the other hand, in the quark sea the total quark density is of order $(k_q^F)^3$, and if the nucleons are composed of quarks their baryon density must be that of the quarks, requiring a suppression of the phase space density of nucleons because $n_N \sim (k_N^F)^3/N_C^3 \sim (k_q^F)^3$. Therefore, the bulk distribution of nucleons with momentum less than the nucleon shell is suppressed.

One can model such a picture in various ways to develop a concrete picture with these features [13]. It is also possible to develop explicit quantum mechanical computations of dual description of quarks and nucleons which exhibit these features [14]. The phenomenological model of McLerran and Reddy proposes that nucleons exist on a shell in momentum space whose thickness is determined so that the density of nucleons is fixed at some value above that of nuclear matter [13]. The Fermi sea of quarks and gluons sit below this shell. In the computations of Fujimoto, McLerran and Tojo, the general features of this model are reproduced in a quantum mechanical computation where nucleons are explicitly composed of quarks, and the occupation numbers of the quark and gluon states are constrained to be between zero and one [14]. The rapid rise in the sound velocity is attributed to the transition to quarkyonic matter that occurs when quark occupation numbers approach one. Similar features are extracted from excluded volume models of nuclear matter, where the overlaps of hard cores may be thought of as corresponding to the density at which the occupation numbers of quarks approach one, and the origin of the hard core repulsive interaction is the Fermi exclusion of filled quark states [15].

The reason the sound velocity increases rapidly in the quarkyonic compared to the nucleonic phase is easily understood. The typical Fermi momentum for a density corresponding to the nuclear matter density is of order Λ_{QCD}. Therefore, the sound velocity squared in nuclear matter is of order $v_s^2 \sim \Lambda_{QCD}^2/M_N^2 \sim 1/N_c^2$. For quarkyonic matter it is $v_s^2 \sim \Lambda_{QCD}^2/M_q^2 \sim 1$, meaning the ratio of sound velocities is of order $1/N_c^2$. The transition occurs at approximately a constant baryon density; therefore, the sound velocity and equation of state change quickly from a soft equation of state to a hard equation of state. In models, this change can sometimes correspond to a peak in the sound velocity [13,14].

To summarize, it is possible to reproduce features of the equation of state inferred from neutron star observations using a quarkyonic description. The rapid change in the sound velocity occurs when the matter becomes quarkyonic. At this point, the degrees of freedom become relativistic, although approximate scale invariance is acquired at higher densities.

There is an issue about how one thinks about the transition to quarkyonic matter. Indeed, such a transition might be thought of as a phase transition to quarks. However, it is not a first- or second-order transition, since the sound velocity does not tend to zero at the transition. Indeed, the sound velocity increases rapidly. In fact, the transition is probably not due to de-confinement, since in the large N_c limit the Debye screening length is

$$r_{Debye}^2 \sim N_c/(k_q^F)^2. \tag{5}$$

This is because Debye screening is induced by a one-loop effect, and at a temperature of zero, such a loop can only be made of quarks, supressing the diagram by a factor of N_c. Therefore, the Debye screening length only becomes of the order of the QCD scale when $k_q^F \sim \sqrt{N_c}\Lambda_{QCD}$. On the other hand, the quarkyonic transition occurs at lower density when $k_q^F \sim \Lambda_{QCD}$. Of course, for $N_c = 3$ it is possible that these transitions are poorly separated.

4. Accelerator Experiments and High-Density Baryonic Matter

One should also ask what can be learned from systematic accelerator studies. One question to address is how one might measure a rapidly varying sound velocity as a function of baryon density at low temperatures. This restricts one to low-energy experiments with low-energy accelerators, such as GSI, FAIR, FRIB or RHIC. Two suggestions have been made.

The first involves measuring fluctuations in the baryon number [16]. With a cumulant of the baryon number distribution defined as

$$\kappa_j =<< (n_B - < n_B >)^j >> . \tag{6}$$

This can be rewritten in terms of thermodynamic quantities as

$$\kappa_j = VT^{j-1}\frac{d^jP}{d\mu_B^j}. \tag{7}$$

where T is the temperature, μ_B the baryon number chemical potential, and P is the pressure. These can be combined together at low temperature to give

$$\frac{d\ ln\ v_s^2}{d\ ln\ n_B} + v_s^2 = 1 - \frac{\kappa_1\kappa_3}{\kappa_2^2}. \tag{8}$$

Another suggested method is to determine the sound velocity transport computations. This is easily performed for vector mean field models, where the sound velocity is directly related to the strength of the mean field [17]. The mean field vector potential is the chemical potential for the baryon number. However,

$$v_s^2 = \frac{n_B}{\mu_B dn_B/d\mu_B} \tag{9}$$

and therefore

$$\mu_B = \mu_B(n_B^0)exp\left(\int_{n_B^0}^{n_B} dn'\ \frac{v_s^2(n')}{n'}\right) \tag{10}$$

such that the effects of the vector field can be known from the dependence of the sound velocity on the density.

While the results of such analyses [16,17] are suggestive that there may be large sound velocities achieved at low densities, our understanding is still in its early stages, and with time this will increase. The results presented here are interesting. The sound velocity squared is first seen to decrease and then approach unity at a density close to that of nuclear matter. This is different from the studies on neutron stars. Neutron stars have an isospin to baryon number of near 1/2 while nuclear matter is closer to zero, so there is no a priori contradiction.

5. Theoretical Issues

Among the theoretical issues, the following are most important:

Can one develop a transport theory method of computations which allows quarkyonic matter to form at high baryon densities in relativistic nuclei collisions? To do this one needs to impose constraints on the nucleon and quark occupation numbers. Perhaps one could simulate nucleonic collisions and then compute the quark content, imposing quark and nucleon occupation numbers between zero and one. This would be sufficient to compute the equation of state for systems that are time-independent. In such a dynamic system, an important issue would be to see how the occupation numbers are distributed and evolve.

Another important question concerns the isospin dependence of the equation of state. For neutron matter, the isospin per nucleon is 1/2, while for nuclear matter it is zero. In principle, the equation of state may be very different for such systems. It might be that the inferred properties from the equation of state of neutron matter is qualitatively different from the inferred collisions of nuclei with low isospin per nucleon.

How does chiral symmetry manifest in quarkyonic matter? Theoretical studies have been carried out that show the pattern of chiral symmetry restoration may be quite arcane [18]. How does one clearly model such matter? Are there consequences at densities found for neutron stars or accelerator collisions?

How does one generalize quarkyonic matter from a temperature of zero to finite density [19]? This is important to describe low-energy heavy-ion collisions and for the collision dynamics of neutron star collisions.

How does one use field theoretical methods, such as mean field theory, to dynamically demonstrate the formation of quarkyonic matter [20,21]?

Funding: This research was funded U.S. DOE under Grant No. DE-FG02-00ER41132.

Data Availability Statement: No data is generated.

Conflicts of Interest: The funders had no role in the design of the study; in the collection, analyses, or interpretation of data; in the writing of the manuscript; or in the decision to publish the results.

References

1. Freedman, B.A.; McLerran, L.D. Fermions and gauge vector mesons at finite temperature and density. III. The ground-state energy of a relativistic quark gas. *Phys. Rev. D* **1977**, *16*, 1169. [CrossRef]
2. Annala, E.; Gorda, T.; Kurkela, A.; Nattila, J.; Vuorinen, A. Evidence for quark-matter cores in massive neutron stars. *Nat. Phys.* **2020**, *16*, 907. [CrossRef]
3. Bedaque, P.; Steiner, A.W. Sound velocity bound and neutron stars. *Phys. Rev. Lett.* **2015**, *114*, 031103. [CrossRef] [PubMed]
4. Fraga, E.S.; Kurkela, A.; Vuorinen, A. Interacting quark matter equation of state for compact stars. *Astrophys. J. Lett.* **2014**, *781*, L25. [CrossRef]
5. Annala, E.; Gorda, T.; Kurkela, A.; Vuorinen, A. Gravitational-wave constraints on the neutron-star-matter Equation of State. *Phys. Rev. Lett.* **2018**, *120*, 172703. [CrossRef]
6. Tews, I.; Carlson, J.; Gandol, S.; Reddy, S. Constraining the speed of sound inside neutron stars with chiral effective field theory interactions and observations. *Astrophys. J.* **2018**, *860*, 149. [CrossRef]
7. Fujimoto, Y.; Fukushima, K.; Murase, K. Extensive studies of the neutron star equation of state from the deep learning inference with the observational data augmentation. *J. High Energy Phys. 2021*, *03*, 273. [CrossRef]
8. Marczenko, M.; McLerran, L.; Redlich, K.; Sasaki, C. Reaching percolation and conformal limits in neutron stars. *Phys. Rev. C* **2023**, *107*, 025802. [CrossRef]

9. Fujimoto, Y.; Fukushima, K.; McLerran, L.D.; Praszalowicz, M. Trace anomaly as signature of conformality in neutron stars. *Phys. Rev. Lett.* **2022**, *129*, 252702. [CrossRef]
10. McLerran, L.; Pisarski, R.D. Phases of dense quarks at large Nc. *Nucl. Phys. A* **2007**, *796*, 83. [CrossRef]
11. Kojo, T. Stiffening of matter in quark-hadron continuity. *Phys. Rev. D* **2021**, *104*, 074005. [CrossRef]
12. Ma, Y.-L.; Rho, M. Manifestation of hidden symmetries in baryonic matter: From finite nuclei to neutron stars. *Mod. Phys. Lett.* **2021**, *36*, 2130012.
13. McLerran, L.; Reddy, S. Quarkyonic matter and neutron stars. *Phys. Rev. Lett.* **2019**, *122*, 122701. [CrossRef] [PubMed]
14. Fujimoto, Y.; McLerran, L.; Kojo, T. Momentum Shell and Rapid Stiffening in Quarkyonic Matter from Explicit Duality: An Exactly Solvable Model. *manuscript in preparation.*
15. Jeong, K.S.; McLerran, L.; Sen, S. Dynamically generated momentum space shell structure of quarkyonic matter via an excluded volume model. *Phys. Rev. C* **2020**, *101*, 035201. [CrossRef]
16. Sorensen, A.; Oliinychenko, D.; Koch, V.; McLerran, L. Speed of sound and baryon cumulants in heavy-ion collisions. *Phys. Rev. Lett.* **2021**, *127*, 042303. [CrossRef]
17. Oliinychenko, D.; Sorensen, A.; Koch, V.; McLerran, L. Measuring the speed of sound using cumulants of baryon number. *arXiv* **2022**, arXiv:2209.04957.
18. Kojo, T.; Hidaka, Y.; Fukushima, K.; McLerran, L.D.; Pisarski, R.D. Interweaving chiral spirals. *Nucl. Phys. A* **2012**, *875*, 94. [CrossRef]
19. Sen, S.; Warrington, N.C. Finite-temperature quarkyonic matter with an excluded volume model. *Nucl. Phys. A* **2021**, *1006*, 122059. [CrossRef]
20. Kumar, A.; Dey, D.; Haque, S.; Mallick, R.; Patra, S.K. Quarkyonic Model for Neutron Star Matter: A Relativistic Mean-Field Approach. *arXiv* **2023**, arXiv:2304.08223.
21. Duarte, D.C.; Hernandez-Ortiz, S.; Jeong, K.S.; McLerran, L.D. Quarkyonic Mean Field Theory. *arXiv* **2023**, arXiv:2302.04781.

symmetry

MDPI

Article

T_{cs} and $T_{c\bar{s}}$ Family Production in Multi-Production Processes

Chuanhui Jiang [1], Yi Jin [2,3,*], Shi-Yuan Li [1,*], Yan-Rui Liu [1] and Zong-Guo Si [1]

[1] School of Physics, Shandong University, Jinan 250100, China; chuanhuijiang@mail.sdu.edu.cn (C.J.); yrliu@sdu.edu.cn (Y.-R.L.); zgsi@sdu.edu.cn (Z.-G.S.)
[2] School of Physics and Technology, University of Jinan, Jinan 250022, China
[3] Guangxi Key Laboratory of Nuclear Physics and Nuclear Technology, Guangxi Normal University, Guilin 541004, China
* Correspondence: ss_jiny@ujn.edu.cn (Y.J.); lishy@sdu.edu.cn (S.-Y.L.)

Abstract: The production mechanism of multiquark exotic hadrons in high energy multiproduction processes lies in the structure of the relevant exotic hadrons as well as in some important aspects of high energy scattering, such as multi-parton interactions, underlying events, etc. At mass pole around 2900 MeV, a family of open charm tetraquarks, T_{cs}s and $T_{c\bar{s}}$s, are observed in B decay. They are also suitable for study in multiproduction processes to obtain more information on their structure. If these resonances are produced as compact four-quark states, one can predict the production properties based on the similarities in their production mechanism to those of Ξ_c, Σ_c, and Λ_c. Physics implies that the colour and baryon number fluctuations of the preconfinement system in high energy scattering can enhance both the baryon and four-quark state production rates via 'diquark fragmentation'. We calculate the production properties of the tetraquark family T_{cs}s and $T_{c\bar{s}}$s at LHC energy for the forthcoming LHC measurements.

Keywords: 2 tetraquark; multi-production; corresponding baryon

1. Introduction

It has long been recognized that multiproduction at high energy is very important to understand various properties of QCD and its confinement mechanism. Its global property is related to the hadronic scattering total cross-section [1,2]. On the other hand, once the production of a specific hadron is considered, the structure of the hadron is also crucially relevant, besides its properties with respect to all the other hadrons. In the quark model of a hadron, mesons and baryons are quark-antiquark hadrons and three-quark hadrons, respectively. (Anti)quarks are bound by strong interactions and they are in colour singlet state. SU(3) colour symmetry also allows clusters of >3 (anti)quarks in colour singlet state. It is possible that these kinds of colour singlet clusters with modest mass are 'real particles', and they are called 'exotic hadrons' [3]. Among all the possibilities, four-quark states ($qq\bar{q}\bar{q}$, tetraquark) and five-quark states ($qqqq\bar{q}$, pentaquark), containing at least one heavy quark (c or b), have been observed in various experiments. Almost all the four- and five-quark states have been observed from the decay of heavier hadrons (e.g., bottom hadrons), rather than being promptly produced from multiproduction processes in high energy scattering. Theoretically, this fact can be understood by that their production rate is generally quite small in multiproduction processes because of the unitarity constraint [4,5], as well as the modest mass of the preconfinement clusters [6,7]. This conclusion is one of the most straightforward deductions from the quark model of a hadron. However, a few exceptions are observed from multiproduction at the LHC, and these exotics contain more heavy quarks and hence cannot be produced from heavier particle (e.g., B meson) decay [8–10]. Frankly, no rule in QCD says that only very heavy exotics can be produced from multi-production, though one may complain that the "trigger" of the perturbative heavy quark production can modify the unitarity constraint of hadronization or that the preconfinement scale of heavy quark clusters can be different from pure light ones [6]. Therefore, it is a crucial goal to study exotic hadrons which are produced *both* from multiproduction and from B decay to understand more details of exotic hadron properties and structures. Recently, a group of tetraquark brothers and sisters of T_{cs} and $T_{c\bar{s}}$ were observed by LHCb from B decay (see Tables 1 and 2 and refs. therein). Their masses varied around 2900 MeV; hence, in this paper we call them the T(2900) family if they are not specified otherwise. It should also be possible to produce this T(2900) family in multiproduction process at the LHC. For the purpose of observing them in multiproduction, it is very important to predict their production rate as well as their kinematic distributions, since the initial state of multiproduction is different from, or rather more complex than, that of the B decay. In particular, without a simple and definite initial state, analysis methods such as partial wave analysis cannot be applied.

To investigate and answer such questions, the key problem is how to describe the production of these tetraquarks, i.e., how these four ingredient quarks are produced and how they become a hadron. This is a very complex problem and the mechanism could be model-dependent, i.e., relevant to the concrete hadronization models [11–13] adopted, which have yet not been well tuned for exotic hadrons. On the other hand, one may rely on other hadrons which have been well measured and hence can be tuned to be well described by a model, more concretely, an event generator. Once the production of such hadrons is fixed, one can deduce the corresponding kinematics for the tetraquark based on the relation of their production mechanism, which is not dependent on concrete hadronization models but only determined by the structure of the tetraquark. The idea is similar to our study on doubly heavy tetraquarks, T_{cc} [14]. In that case, the relation to Ξ_{cc} was mainly determined by the cc doubly heavy diquark. In the present case, we can refer to Λ_c, Σ_c and Ξ_c for the study of the kinematic spectra as well as production rate of the T(2900) family. The relations are mainly determined by the diquark which is constructed by a charm quark with a light quark. To explain this point, we again start from the colour structure of the tetraquark and demonstrate its relation with the above 'corresponding baryons'.

It is a common property of all kinds of multiquark hadrons that the bound (anti)quarks inside a multiquark hadron can be grouped into several clusters. The ways of grouping these (anti)quarks are not unique, as it is simply known from group theory that the reduction of a direct product of several representations is not unique. In some grouping ways, each cluster can be a colour singlet, indicating a picture of 'hadron molecule'. However, such a way of clustering the (anti)quarks is not necessary, since the only requirement is that the *whole* set of these clusters are a colour singlet. For example, the system $q_1\bar{q}_2q_3\bar{q}_4$ (the constituents of a four-quark state) can be decomposed/clustered in the following various ways:

$$(q_1q_3)_{\bar{3}} \otimes (\bar{q}_2\bar{q}_4)_3 \to 1 \tag{1}$$

$$(q_1\bar{q}_2)_{1\ or\ 8} \otimes (q_3\bar{q}_4)_{1\ or\ 8} \to 1 \tag{2}$$

$$\cdots$$

In the above examples, only the second case, when the two $q\bar{q}$ pairs are both colour singlets, could it potentially be considered as a hadron molecule. Group theory analysis here is applicable to the quark states as well as to the quark field operators [15]. If we go beyond such a static picture of the structure of multiquark hadrons, considering the situation more dynamically, the colour interactions in the system via exchanging gluons can change the colour state of each individual cluster, so each kind of grouping/reduction can change to the others and seems to have no special physical cause. Such an ambiguity has been considered in many hadronization and decay processes in the framework of 'colour recombination/rearrangement' [16–18]. The above fact makes the possibility of considering the multiquark hadron structure in a unique and uniform way difficult, while it leads to the possibility of introducing some phenomenological duality. It is easy to find that it is possible for multiquark hadrons to produce as 'hadron molecule formation', but the subsequent colour interactions in the system can eventually transit this 'molecule' into a 'real' (compact) multiquark hadron, at least by some probability, and vice versa. Hence, as is argued in [14], this ambiguity just shows that in cases where various kinds of model explain the data of the static properties well, such as masses and decay widths, and we cannot determine the structure from the static properties (a 'compact' multiquark state or a hadron molecule), we can still try to investigate the production mechanism of the T(2900) family relying on only their quark ingredients (necessarily requiring four quarks) and leaving out the details of the structure. If there are significant differences in the production properties among the T(2900) family members observed in the multiproduction processes, these signals are for sure a sound implication of their structure details. In this case, this family can be taken as an excellent example for the study of the production mechanism to gain insight into the exotic hadronic structure.

The T(2900) family has been long studied theoretically (for a theoretical review, see [19] and refs. therein). All T(2900) family members are listed in Tables 1 and 2 (with references). Some have been observed from B decay by the LHCb collaboration.

Table 1. Considered and observed tetraquark states for the *P* even case. The subscript *J* denotes the spin of the state: 0, 1, or 2.

Contents	States	Observed	Mass	Width
$cs\bar{u}\bar{u}$	T^{a-}_{csJ}			
$cs\bar{u}\bar{d}$	T^{a0}_{csJ}, T^{f0}_{csJ}	$T_{cs0}(2900)^0$ [20,21]	$2.866 \pm 0.007 \pm 0.002$ GeV	$57 \pm 12 \pm 4$ MeV
$cs\bar{d}\bar{d}$	T^{a+}_{csJ}			
$c\bar{s}u\bar{d}$	$T^{a++}_{c\bar{s}J}$	$T^{a}_{c\bar{s}0}(2900)^{++}$ [22,23]	$2.908 \pm 0.011 \pm 0.020$ GeV	$0.136 \pm 0.023 \pm 0.011$ GeV
$c\bar{s}u\bar{u}$, $c\bar{s}d\bar{d}$	$T^{a+}_{c\bar{s}J}$, $T^{f+}_{c\bar{s}J}$			
$c\bar{s}d\bar{u}$	$T^{a0}_{c\bar{s}J}$	$T^{a}_{c\bar{s}0}(2900)^{0}$ [22,23]	$2.908 \pm 0.011 \pm 0.020$ GeV	$0.136 \pm 0.023 \pm 0.011$ GeV

Table 2. Considered and observed tetraquark states for the *P* odd case. The subscript *J* denotes the spin of the state: 0, 1, 2, or 3.

Contents	States	Observed	Mass	Width
$cs\bar{u}\bar{u}$	$T^{\pi-}_{csJ}$			
$cs\bar{u}\bar{d}$	$T^{\pi 0}_{csJ}$, $T^{\eta 0}_{csJ}$	$T_{cs1}(2900)^0$ [20,21]	$2.904 \pm 0.005 \pm 0.001$ GeV	$110 \pm 11 \pm 4$ MeV
$cs\bar{d}\bar{d}$	$T^{\pi+}_{csJ}$			
$c\bar{s}u\bar{d}$	$T^{\pi++}_{c\bar{s}J}$			
$c\bar{s}u\bar{u}$, $c\bar{s}d\bar{d}$	$T^{\pi+}_{c\bar{s}J}$, $T^{\eta+}_{c\bar{s}J}$			
$c\bar{s}d\bar{u}$	$T^{\pi 0}_{c\bar{s}J}$			

From the content of the quark components and considering their colour structure ambiguity, as discussed in the above paragraphs, we can conjecture the following correspondence between the T(2900) family and the charm hyperon for the study of the production in multiproduction processes:

(1) For the T_{cs} brothers and sisters, their production can be treated as the hadronization of the cs diquark and hence can be related to Ξ_c. The difference between a cs diquark transiting to a Ξ_c or a T_{cs} is that this diquark combines a light quark (u or d) or a light anti-diquark ($\bar{u}\bar{d}$, $\bar{u}\bar{u}$, or $\bar{d}\bar{d}$).

(2) According to a similar consideration, $T_{c\bar{s}}$ brothers and sisters can be considered as the fragmentation of a cq ($q = u$ or d, the same in the following) diquark. Once it combines with a $\bar{s}\bar{q}$ anti-diquark, it becomes one $T_{c\bar{s}}$, while it will become a Λ_c or Σ_c baryon by combining a light quark u or d.

We adopt the 'partially breaking $SU_f(3)$ symmetry' in hadronization, with the production ratio of the quark flavours created from vacuum by strong interactions as $u : d :$

$s = 1 : 1 : \lambda$. At the same time, the light diquark production rate has a suppression factor ζ [12]. By fixing these ratios with the above 'correspondence', we can just fully employ the production properties of Λ_c, Σ_c, and Ξ_c to predict the production of the T(2900) family.

2. Kinematic Spectra of Corresponding Baryons and Production of T(2900) Family at the LHC

As is mentioned in last section, the $T_{cs}/T_{c\bar{s}}$ four-quark states can be taken as the combination of a cs/cq cluster (diquark) with a light anti-diquark $\bar{q}\bar{q}'/\bar{s}\bar{q}'$ produced from strong interactions. In high energy processes such as proton–proton collisions at the LHC, charm quarks are produced from hard scattering (including the semi-hard multiparton interactions) and then evolve to softer scales via radiations. Before the hadronization process, preconfinement [7] clusters form, which contain charm quarks as well as light quarks produced during the evolution radiation and other softer strong interactions. The preconfined system is the end of perturbative QCD (PQCD) evolution, with the formation of colour singlet clusters. These clusters have modest masses independent from the centre-of-mass energy of the scattering process. Each colour singlet cluster will independently hadronize into hadrons. There is much evidecne for preconfinement cluster formation in high energy processes, one of which is the local parton hadron duality (for details, see [6] and refs. therein). It has long been argued that the colour structure of the preconfined system is not unique [17,18]. It has also been recognized that different colour structures will lead to different non-trivial baryon number distributions of the colour singlet clusters, which is referred to as the baryon number fluctuation of the preconfined system [6]. One of the colour structures (Equation (1)) favours the production of both charm hyperons and tetraquarks. In concrete calculations, the proton–proton scattering occuring at very high energy as the LHC suffers complexity of initial states, multi-parton scattering, and underlying events. It is impossible to employ simple analytical formulations to describe such complex processes with the special colour connections. Thus, we have to employ event generators as a practical framework to perform the calculation. Though most event generators can help to describe the initial states, multi-parton scattering, and underlying events (we will not repeat the details of the models here), we still have to solve two linked difficulties: The first is that no event generator has been sufficiently well developed to describe exotic hadrons. This difficulty, specified in our concrete problem, can be solved by the idea introduced in the above section, i.e., to employ the 'corresponding' baryons as a calibration to describe the T(2900) family. However, this idea at once encounters another difficulty, i.e., the generators fail to describe the baryon production data well. The calculated results are much lower than the data [24]. Therefore, here we must first try to find ways to tune the generator according to the baryon production data. The key point is to find the physics which is missing for the production of baryons (hence also of tetraquarks).

For the feasibility of the discussion, let us first take Ξ_c production as an example. Here, we employ the PYTHIA generator with the Lund string 'fragmentation' model [12,25] as the basic framework to calculate the differential cross-section. Its results are much lower than the experimental data [24], and we must add the missing contributions. As demonstrated in [26], a higher twist contribution which we refer to as 'combination' has to be taken into account and added to the lowest twist 'fragmentation', with the latter corresponding to the string fragmentation in the PYTHIA. As described in [26], a cs diquark combined with a light quark produced from the underlying events can be described by the following inclusive differential cross-section,

$$2E\frac{d\sigma}{d^3K} = \sum_{ab}\int dx_1 dx_2 f_1^a(x_1) f_2^b(x_2) \frac{d\hat{\sigma}_{ab}}{d\mathcal{I}} \frac{1}{\xi^2} \frac{(2\pi)^2}{(2M)^2} P(\xi_l)\tilde{F}(\xi,\xi_l)|_{\xi+\xi_l=1}. \quad (3)$$

In the above equation, $d\mathcal{I}$ is the dimensionless invariant phase space for the 'two-body' partonic final state $cs + x$, where x is treated as one particle. At present, the relevant matrix elements describing the combination contribution are yet unknown. Motivated by the results of [26], we can parameterize the combination contribution according to its exponential behaviour and add it to the PYTHIA result:

$$\left(\frac{d^2\sigma}{dydp_T}\right)_{mod} = \left(\frac{d^2\sigma}{dydp_T}\right)_{pythia} \cdot a + \alpha e^{-\beta p_T}. \tag{4}$$

By fitting the best LHC experimental data, we obtain $a = 1.000$, $\alpha = 401.574$ µb GeV^{-1}c, and $\beta = 0.961$ (GeV/c)$^{-1}$, respectively. The result can be seen in Figure 1, and we obtain a good description of the data by adding the 'combination contribution'. Here, we would like to take the chance to clarify the terminology 'combination' in various circumstances:

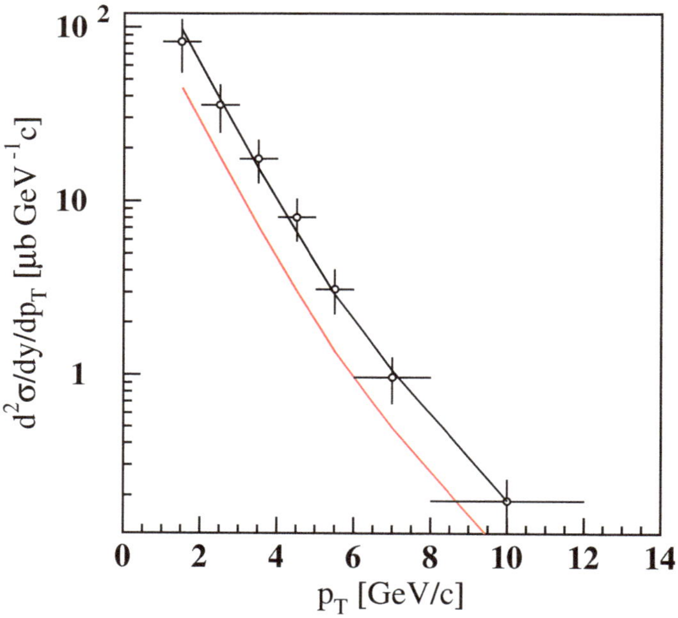

Figure 1. Cross-section of the prompt Ξ_c^0 as a function of p_T in pp collisions at $\sqrt{s}$ = 13 TeV. The black circles are the experimental data [24], while the black solid line is for Equation (4) with the fitted parameters. The red solid line is for the prediction of the distribution of the T_{cs}. The isospin and spin symmetry is adopted. We have averaged among all the isospin and spin states.

(1) In the case of PQCD factorization, this terminology refers to a higher twist contribution described by some matrix elements and can be added with the lowest twist 'fragmentation' contribution, as in [26]. At high energies, such as in the LHC, the underlying events highly contribute to the production of soft quarks; the combination contribution can hence be largely enhanced. The fitted results shown above are consistent with this consideration. Correspondingly, the result calculated by PYTHIA is only the lowest twist fragmentation contribution, which is described by the string fragmentation model, and one can find the fragmentation function is effectively that in [25]:

$$f(z) \propto z^{-1}(1-z)^d exp(-bm_\perp^2/z), \tag{5}$$

where d and b are free parameters. In our simulation program, we take $d = 0.3$ GeV^{-2} and $b = 0.58$ GeV^{-2}, as used in PYTHIA [25]. This part of the contribution is tuned with the e^+e^- data, etc.

(2) In the case of a hadronization model, combination refers to an intuitive picture to explain and describe the hadronization process. There are various quark combination models to describe hadronization (see [11] and references therein). However, introducing the production mechanism of exotic multiquark hadrons is a tough problem, in which the unitarity is a difficulty. One has to take a comprehensive account of all exotics. Here, to make it feasible to only study the T(2900) family inclusively, we choose the PYTHIA rather than the combination models. However, except in the fitting of the prompt charm hyperon by the above Formula (3), we refer to 'combination' always by the second meaning.

Similarly, we also calculate the differential cross-section of the prompt Λ_c^+ (there are not enough Σ_c data, so we neglect the discussion of it in the following), and obtain the corresponding equation as (4), while the parameters are $a = 0.214$, $\alpha = 2454.609$ μb GeV^{-1}c, and $\beta = 0.928$ (GeV/c)$^{-1}$, respectively. The results are shown in Figure 2.

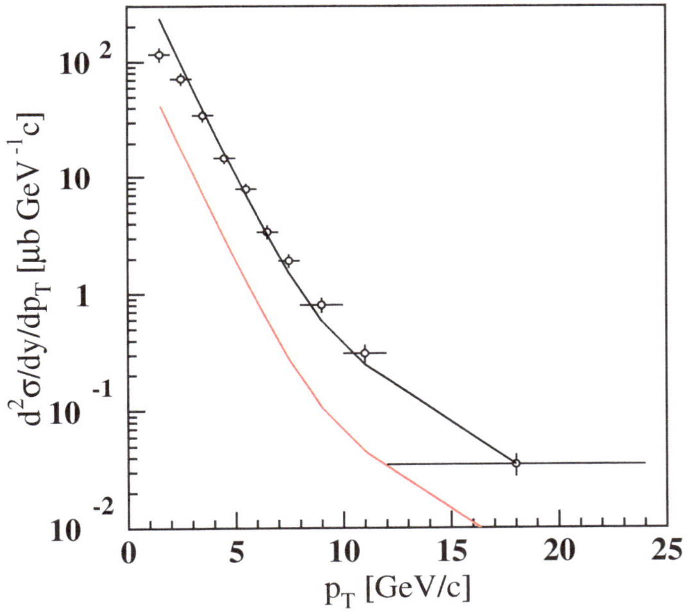

Figure 2. Cross-section of the prompt Λ_c^+ as a function of p_T in pp collisions at $\sqrt{s}$ = 13 TeV. The black circles are experimental data [27], while the black solid line is for Equation (4) with the fitted parameters. The red solid line is for the prediction of the distribution of the $T_{c\bar{s}}$. Isospin and spin symmetry are adopted. We have averaged among all these isospin and spin states.

With the charm hyperon differential cross-sections in hand, we can predict the tetraquark production. The similarity and difference between the charm hyperons, Ξ_c and Λ_c, and the tetraquark hadrons, T_{cs} and $T_{c\bar{s}}$, in their kinematic spectra lie in many aspects; here we only take the most simple model with the help of the Peterson formula [28]:

$$f(z) \propto \frac{1}{z(1 - 1/z - \epsilon_Q/(1-z))^2}, \tag{6}$$

where z is defined by $p_+^{\text{hadron}}/p_+^{\text{cluster}}$, with p_+ being the sum of the energy and the momentum projected in the moving direction of the cluster which hadronize to the correponding hadron, and the parameter ϵ_Q is expected to scale between flavours. In practice, different ϵ_Q values inversely proportional to the hadron square masses are used for tetraquarks T_{cs} and $T_{c\bar{s}}$ with respect to charm hyperons Ξ_c and Λ_c, respectively. The sampling of the tetraquark momenta hence is determined by the fitted charm hyperon momentum distribution and the above relative relation.

For the total production rate, we adopt the most general ansatz for production, the breaking of $SU_f(3)$ flavour symmetry, as mentioned in the Introduction. For the production of quark flavour, the ratio is $u : d : s = 1 : 1 : \lambda$, with $\lambda \sim 0.3$ for a vacuum. The quark to diquark production ratio is $1 : \zeta$, with ζ varying (increasing) with multiplicity [6]. For the multiproduction process of pp collision at high energies of up to 13 TeV at the LHC, the multiplicity is large, so we can take $\zeta \sim \lambda$ for simplicity. Based on this consideration, the T_{cs} has an extra suppression factor $\zeta \sim \lambda$ with respect to Ξ_c when sampling in the generator programme, while $T_{c\bar{s}}$ gets an extra suppression factor $\zeta \times \lambda \sim \lambda^2$. The results of T_{cs} and $T_{c\bar{s}}$ production at the LHC are displayed in the figures with red solid lines.

3. Conclusions and Discussion

In this paper, we study the T(2900) tetraquark family in multiproduction processes at the LHC. We first investigate the production mechanism of this kind of hadron, emphasizing its similarity with the charm hyperon production mechanism. By this similarity, we predict the production rate as well as the kinematics spectra. This is crucial information for the forthcoming relevant measurements at the LHC. Based on our prediction, this family can be measured at the LHC, since the corresponding charm hyperons have been well measured. The production rate of each one of the T(2900) family is only several times smaller than that of the relevant charm hyperon, and the kinematic spectra are almost the same. Furthermore, various decay channels have been carefully studied (see, e.g., [29,30]), and two body decays are dominant. A reconstruction of the tetraquark hadrons is practical.

In the above sections, to find the relation between charm hyperons and tetraquarks, our study is based on the concept of colour reconnection, as well as the intuitive physical picture of hadronization models. Prof. Qu-Bing Xie is one of the pioneers who initialized these ingenious concepts [11,12,17,18]. To further prove the validity of these physical pictures on the study of T(2900) family production, we propose that the light hadrons produced adjointly with the charm hyperon or T(2900) can provide additional clues via string effects (see a likely example in [14]). This measurement can be performed after the T(2900) tetraquark hadron(s) has (have) been observed.

In the analysis and calculations of this paper, we did not rely on the detailed structure of the T(2900) family, nor on that of the corresponding baryons, e.g., information on their spins, isospin/$SU_f(3)$ wave functions, or colour structures. One of the reasons is of course the lack of knowledge on the tetraquark structure. However, for our present purpose, it is more practical to ignore the details of the structure and give a 'baseline' prediction for the production of tetraquark hadrons via the best-fitted baryon production cross-sections. By comparing our prediction with the forthcoming data, one can take further steps to model the structure of the tetraquark more precisely. This is in fact a very general way to study the structure of hadrons with the help of the multiproduction processes, and we believe that a deeper understanding of the hadron structure and confinement mechanism lies in the comprehensive study not only of mesons and baryons but also of various exotic hadrons. This belief and ideas, as we know, were among the original motivations of Profs. Lian-Shou Liu, Qu-Bing Xie, and their collaborators when they began to study hadronzation and multiparticle dynamics (according to private conversations and their group seminar talks).

Author Contributions: Conceptualization, S.-Y.L.; Methodology, Y.J., Y.-R.L. and Z.-G.S.; Software, C.J. All authors have read and agreed to the published version of the manuscript.

Funding: This work was supported in part by the National Natural Science Foundation of China (NSFC) (grant nos. 12275157, 12235008, and 11775130). Y.J. is supported by the Open Project of Guangxi Key Laboratory of Nuclear Physics and Nuclear Technology under grant no. NLK2021-07.

Data Availability Statement: Not applicable.

Acknowledgments: We thank all the members of our particle theory group for discussions and collaborations.

Conflicts of Interest: The authors declare no conflict of interest.

References

1. Cheng, H.; Wu, T.T. *Expanding Protons: Scattering at High-Energies*; MIT Press: Cambridge, MA, USA, 1987.
2. Donnachie, A.; Landshoff, P.V. Total cross sections. *Phys. Lett. B* **1992**, *296*, 227–232.
3. Gershon, T. Exotic hadron naming convention. *arXiv* **2022**, arXiv:2206.15233. [CrossRef]
4. Han, W.; Li, S.Y.; Shang, Y.H.; Shao, F.L.; Yao, T. Exotic hadron production in quark combination model. *Phys. Rev. C* **2009**, *80*, 035202.
5. Jin, Y.; Li, S.Y.; Liu, Y.R.; Meng, L.; Si, Z.G.; Zhang, X.F. Exotic hadron bound state production at hadron colliders. *Chin. Phys. C* **2017**, *41*, 083106. [CrossRef]
6. Li, S.Y. Baryon Number Fluctuation of Preconfinement Clusters and Transverse Momentum: Dependence of Bottom Baryon Production. In Proceedings of the Moriond QCD 2019, La Thuile, Italy, 23–30 March 2019; pp. 107–110. [CrossRef]
7. Amati, D.; Veneziano, G. Preconfinement as a Property of Perturbative QCD. *Phys. Lett. B* **1979**, *83*, 87–92.
8. LHCb Collaboration. Observation of structure in the J/ψ -pair mass spectrum. *Sci. Bull.* **2020**, *65*, 1983–1993. [CrossRef]
9. LHCb Collaboration. Observation of an exotic narrow doubly charmed tetraquark. *arXiv* **2021**, arXiv:2109.01038. [CrossRef] [PubMed]
10. LHCb Collaboration. Study of the doubly charmed tetraquark T^+_{cc}. *arXiv* **2021**, arXiv:2109.01056.
11. Xie, Q.B.; Liu, X.M. Quark Production Rule in $e^+e^- \to$ Two Jets. *Phys. Rev. D* **1988**, *38*, 2169–2177.
12. Andersson, B.; Gustafson, G.; Ingelman, G.; Sjostrand, T. Parton Fragmentation and String Dynamics. *Phys. Rept.* **1983**, *97*, 31–145. [CrossRef]
13. Corcella, G.; Knowles, I.G.; Marchesini, G.; Moretti, S.; Odagiri, K.; Richardson, P.; Seymour, M.H.; Webber, B.R. HERWIG 6: An Event generator for hadron emission reactions with interfering gluons (including supersymmetric processes). *JHEP* **2001**, *01*, 010. [CrossRef]
14. Jin, Y.; Li, S.Y.; Liu, Y.R.; Qin, Q.; Si, Z.G.; Yu, F.S. Color and baryon number fluctuation of preconfinement system in production process and Tcc structure. *Phys. Rev. D* **2021**, *104*, 114009. [CrossRef]
15. Anikin, I.V.; Pire, B.; Teryaev, O.V. Search for isotensor exotic meson and twist 4 contribution to $\gamma^*\gamma \to \rho\rho$. *Phys. Lett. B* **2005**, *626*, 86. [CrossRef]
16. Han, W.; Li, S.Y.; Si, Z.G.; Yang, Z.J. Colour connection and diquark fragmentation in $\mathrm{e}^+\mathrm{e}^- \to q\_1\bar{q}\_2q\_3\bar{q}\_4 \to$ h's process. *Phys. Lett. B* **2006**, *642*, 62–67. [CrossRef]
17. Si, Z.G.; Wang, Q.; Xie, Q.B. Probability of color rearrangement at partonic level in hadronic W+ W- decays. *Phys. Lett. B* **1997**, *401*, 107. [CrossRef]
18. Gustafson, G.; Hakkinen, J. Color interference and confinement effects in W pair production. *Z. Phys. C* **1994**, *64*, 659. [CrossRef]
19. Chen, H.X.; Chen, W.; Liu, X.; Liu, Y.R.; Zhu, S.L. An updated review of the new hadron states. *Rep. Prog. Phys.* **2023**, *86*, 026201. [CrossRef]
20. LHCb Collaboration. A model-independent study of resonant structure in $B^+ \to D^+D^-K^+$ decays. *Phys. Rev. Lett.* **2020**, *125*, 242001. [CrossRef]
21. LHCb Collaboration. Amplitude analysis of the $B^+ \to D^+D^-K^+$ decay. *Phys. Rev. D* **2020**, *102*, 112003. [CrossRef]
22. LHCb Collaboration. First observation of a doubly charged tetraquark and its neutral partner. *arXiv* **2022**, arXiv:2212.02716.
23. LHCb Collaboration. Amplitude analysis of $B^0 \to \overline{D}^0 D_s^+ \pi^-$ and $B^+ \to D^- D_s^+ \pi^+$ decays. *arXiv* **2022**, arXiv:2212.02717.
24. ALICE Collaboration. Measurement of the Cross Sections of Ξ_c^0 and Ξ_c^+ Baryons and of the Branching-Fraction Ratio BR($\Xi_c^0 \to \Xi^- e^+ \nu_e)/\mathrm{BR}(\Xi_c^0 \to \Xi^- \pi^+)$ in pp collisions at 13 TeV. *Phys. Rev. Lett.* **2021**, *127*, 272001.
25. Sjötrand, T.; Eden(Nordita), P; Friberg, C.; Lonnblad, L.; Miu, G.; Mrenna, S.; Norrbin, E. High-energy physics event generation with PYTHIA 6.1. *Comput. Phys. Commun.* **2001**, *135*, 238. [CrossRef] [PubMed]
26. Jiang, C.; Li, H.; Li, S.Y.; Liu, S.; Yin, X. D^* jet fragmentation function paradox and combination mechanism. *Commun. Theor. Phys.* **2023**, *75*, 025201. [CrossRef]
27. ALICE Collaboration. Measurement of Prompt D^0, Λ_c^+, and $\Sigma_c^{0,++}$(2455) Production in Proton–Proton Collisions at $\sqrt{s}$ = 13 TeV. *Phys. Rev. Lett.* **2022**, *128*, 012001. [CrossRef]
28. Peterson, C.; Schlatter, D.; Schmitt, I.; Zerwas, P.M. Scaling Violations in Inclusive e+ e- Annihilation Spectra. *Phys. Rev. D* **1983**, *27*, 105. [CrossRef]
29. Liu, F.X.; Ni, R.H.; Zhong, X.H.; Zhao, Q. Charmed-strange tetraquarks and their decays in the potential quark model. *arXiv* **2022**, arXiv:2211.01711. [CrossRef]
30. Lian, D.K.; Chen, W.; Chen, H.X.; Dai, L.Y.; Steele, T.G. Strong decays of $T_{cs0}(2900)^{++/0}$ as a fully open-flavor tetraquark state. *arXiv* **2023**, arXiv:2302.01167.

symmetry MDPI

Article

QCD Matter and Phase Transitions under Extreme Conditions †

Mei Huang [1,*,‡] and Pengfei Zhuang [2,*,‡]

1 School of Nuclear Science and Technology, University of Chinese Academy of Sciences, Beijing 100049, China
2 Department of Physics, Tsinghua University, Beijing 100084, China
* Correspondence: huangmei@ucas.ac.cn (M.H.); zhuangpf@tsinghua.edu.cn (P.Z.)
† This work is in memory of Professor Lianshou Liu.
‡ These authors contributed equally to this work.

Abstract: The interplay of chiral dynamics and gluodynamics plays an essential role in the nonperturbative QCD region, and the chiral phase transition and deconfinement phase transition are the main topics of QCD phase transitions under extreme conditions, e.g., finite temperature and/or baryon density, strong magnetic fields, and fast rotation. We briefly introduce our own recent progress from the theoretical side on QCD phase transitions and phase diagrams under extreme conditions related to the early universe, compact stars, and heavy-ion collisions.

Keywords: QCD phase transitions; chiral dynamics; gluodynamics

1. Introduction

Quarks and gluons are fundamental building blocks of visible matter, and their interaction is described by quantum chromodynamics (QCD) [1,2]. QCD matter (here we mean quark matter) (Although nuclear matter should be considered as one type of QCD matter, in this paper, in most cases, QCD matter refers to matter consisting of quarks and gluons, i.e., quark matter) existed at around 10 microseconds in the early universe [3] and exists inside the core of compact stars [4,5], and it can be created through relativistic heavy-ion collisions [6]. Thus, understanding quark matter's properties and QCD phase transitions under extreme conditions is of special interest in investigating the evolution of the universe, exploring the structure of compact stars, and understanding observable heavy-ion collisions [7,8].

The interplay of chiral dynamics and gluodynamics plays an essential role in the nonperturbative QCD region and determines the QCD phase structures. The chiral phase transition and deconfinement phase transitions are two important phase transitions well defined in two extreme quark mass limits, respectively. The chiral phase transition is well defined in the chiral limit characterized by the order parameter of the chiral condensation of light flavors $\langle \bar{\psi}\psi \rangle$ and the deconfinement phase transition, which is characterized by the order parameter of the Polyakov-loop $\langle L \rangle$ in the heavy quark mass limit.

At high temperatures, QCD phase transitions can be calculated by using a lattice. As shown in Figure 1 taken from [9], the QCD phase transitions are flavor- and quark-mass-dependent [7–10]. For the three-flavor massless quark limit or three light quarks, the chiral phase transition is of first-order at the critical temperature around $T_c^{\chi} = 155\,\text{MeV}$, and in the case of a two-flavor degenerate massless u, d quark and a relative heavy s quark, the chiral phase transition is of second-order around the critical temperature $T_c^{\chi} = 175\,\text{MeV}$; when the u, d quark has a small current quark mass, the chiral phase transition will become a cross-over. The lattice calculation [11] shows that, with the physical quark mass, the chiral phase transition at finite temperature in the hot early universe is not a real phase transition, but a smooth cross-over. When the quark mass is heavy and becomes infinity, QCD becomes the pure gauge theory of $SU(3)_c$, and lattice gives the first-order deconfinement phase transition from the Polyakov-loop-order parameter at the critical temperature around $T_c^d = 270$ MeV. However, it is not known how the chiral dynamics interplay with gluodynamics and affect the phase transitions.

The properties of dense QCD matter are important to understand the structure of compact stars, which is related to heavy-ion collisions at low collision energy; see the review article [6]. The lattice calculation is still facing the sign problem at finite baryon chemical potential, and some progress has been made in exploring the phase transitions by using the Taylor expansion of μ_B/T [12–15] and using the simulation of imaginary chemical potential(s) [16]. Through symmetry analysis [17,18] and predictions from effective chiral models, e.g., the Nambu–Jona–Lasinio (NJL) model [19,20] and its Polyakov-loop improved version [21–23], the Polyakov-loop linear sigma model and hadron resonance gas model [24–26], and the Dyson–Schwinger equations (DSEs), which considers quarks and gluons' degrees of freedom [27,28], as well as the five-dimensional holographic QCD model [29], which can take into account chiral dynamics and gluodynamics, etc., it is believed that, in the case of two flavors, the chiral phase transition becomes first order at high baryon chemical potential, and the end point along the first-order phase transition line is called the critical end point (CEP). The search for the QCD CEP is one of the main goals of the Beam Energy Scan (BES) program for relativistic heavy-ion collisions (RHICs), the future program at the Facility for Antiproton and Ion Research (FAIR) at Darmstadt, Germany, and the Nuclotron-based Ion Collider Facility (NICA) at JINR, Dubna.

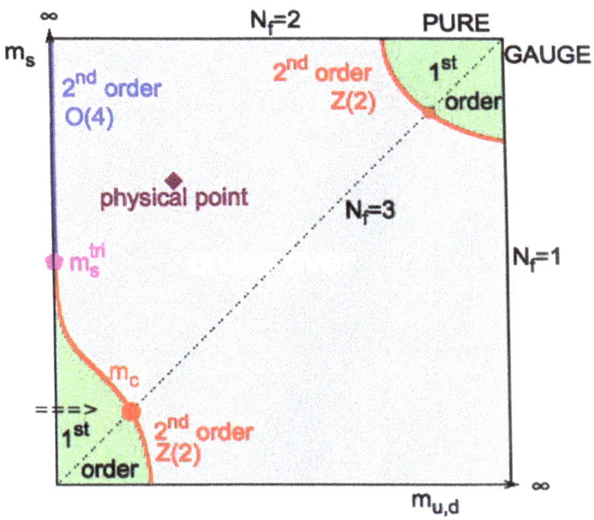

Figure 1. Lattice results on QCD phase diagram at high temperature with a degenerate u, d quark with $m_u = m_d$ and a strange quark with mass m_s. The figure is taken from [9].

Through non-central heavy-ion collisions, the strong magnetic field with the magnitude of 10^{18-19} Gauss and angular momentum of QGP at a range of 10^4–$10^5\hbar$ can be generated [30–35]. Both the magnetic field and angular momentum will polarize the medium, thus changing the properties of QCD matter.

In this short overview, we summarize our own contribution from the theoretical side on the recent progress of QCD phase transitions and QCD matter properties under extreme conditions related to the early universe, compact stars, and heavy-ion collisions.

2. Imprints of QCD Epoch in the Early Universe

The observation of GW signals by LIGO and the Virgo Collaboration [36] 100 years after their prediction [37,38] opens an exciting era for cosmology. Except that the GWs ca be produced through compact binary merges and inspirals, it has also been predicted that the primordial stochastic GW could be produced in the very early stages of the universe through a strong first-order phase transition (FOPT) during inflation, Grand Unified Theory (GUT) symmetry breaking, electroweak phase transition (EWPT) [39–46], as well as QCD phase transition [3,47–50]. GWs carry unique imprints of the early universe about these phase transitions on the cosmic microwave background [51–54]. Except the GWs, another interesting topic during the QCD epoch is the generation of the primordial black holes (PBHs) predicted 50 years ago [55–57], which are possible candidates for dark matter. In the QCD phase transitions around the temperature of $T{\sim}200$ MeV, one would expect the PBH mass to show a peak in the solar mass range [58–60].

In the Minimal Standard Model (SM), the EWPT is well known as a cross-over [39], and it can be a strong FOPT [61] when new physics is considered such as the singlet extension of the SM and two-Higgs-doublet models [40–46]. Similarly, though the lattice QCD predicted that, for a physical mass with $m_u = m_d << m_s$, the system would experience a cross-over in the QCD epoch, there are still other physical conditions in the QCD epoch, and the system experiences a strong FOPT. For example, the pure gluon system might appear in the temperature region below the EWPT, and it exhibits a strong FOPT, as shown by the lattice results [10], as well as the holographic QCD description [62–64], and possible glueball dark matter was investigated in [65,66]. Another interesting FOPT in the QCD epoch is related to the sphaleron transition, which induces chirality imbalance and the violation of $\mathcal{P}$- and $\mathcal{CP}$-symmetry, and the gravitation wave and primordial black hole induced by chirality imbalance under a magnetic field were investigated in [67].

QCD has an important $U(1)_A$ axial symmetry, which is anomalously broken by quantum effects, and the $U(1)_A$ problem or the puzzle of the η and η' mass difference was resolved by the instantons proposed by 't Hooft [68,69]. The chiral anomaly is closely related to the non-trivial topological QCD θ vacuum structure characterized by the integer Chern–Simons number $N_{cs} = \frac{g^2}{32\pi^2}\int \mathrm{d}^4x\, F^a_{\mu\nu}\tilde{F}^{\mu\nu}_a$ [70]. Different Chern–Simons sectors are connected through instanton and sphaleron transitions at zero and finite temperatures, respectively. The Chern–Simons number change through the axial anomaly induces the chirality imbalance between the right-handed and left-handed quarks, $N_5 = N_R - N_L = \int d^4x \partial_\mu j^\mu_5 = -2N_f N_{cs}$, represented by an axial chemical potential μ_5, and results in a violation of $\mathcal{P}$- and $\mathcal{CP}$-symmetry [71–74]. It has been proposed in non-central heavy-ion collisions with a strong magnetic field that this chirality imbalance and local $\mathcal{P}$- and $\mathcal{CP}$-symmetry violation might produce the chiral magnetic effect (CME) [75–77], i.e., parity-odd domains lead to the charge separation of quarks, which has been one of the main goals for RHICs.

Based on the instanton–anti-instanton molecule picture [78–80], in [81,82], it was wn that an effective axial chemical potential μ_5, which describes the chirality imbalance, e dynamically induced by a repulsive axial vector interaction. The Lagrangian density he form:

$$= \bar{\psi}i\gamma_\mu D^\mu\psi + G_S\left[(\bar{\psi}\psi)^2 + \left(\bar{\psi}i\gamma^5\tau\psi\right)^2\right] - G_V(\bar{\psi}\gamma^\mu\psi)^2 - G_A\left(\bar{\psi}\gamma^\mu\gamma^5\psi\right)^2.$$

ne Pauli matrices in the flavor space and G_S, G_V, and G_A are the coupling he scalar, the vector, and the axial vector channels. Quarks with electric e to a magnetic field $\vec{B} = (0,0,B)$ with $A_\mu = (0,0,-xB,0)$ through the ive $D_\mu = \partial_\mu - iq_f A_\mu$. Introducing the quark condensate $\sigma = -2G_S\langle\psi\bar{\psi}\rangle$, der parameter of the chiral phase transition and the dynamical chiral $_5 = -2G_A\langle\bar{\psi}\gamma^0\gamma^5\psi\rangle$ describing the chirality imbalance between left-rks, one can derive the thermodynamical potential:

$$
\begin{aligned}
\Omega = \frac{\sigma^2}{4G_S} - \frac{\mu_5^2}{4G_A} - N_c \sum_{u,d} \frac{|q_f B|}{2\pi} \sum_{s,k} \alpha_{sk} \int_{-\infty}^{+\infty} \frac{\mathrm{d}p_z}{2\pi} f_\Lambda^2(p)\omega_{sk}(p) \\
- 2N_c T \sum_{u,d} \frac{|q_f B|}{2\pi} \sum_{s,k} \alpha_{sk} \int_{-\infty}^{+\infty} \frac{\mathrm{d}p_z}{2\pi} \ln\left(1 + e^{-\beta\omega_{sk}(p)}\right).
\end{aligned} \tag{1}
$$

where the eigenvalues of the Dirac operator are $\omega_{sk} = \sqrt{\sigma^2 + (p + s\mu_5 \mathrm{sgn}(p_z))^2}$ and the spin degeneracy factor is $\alpha_{sk} = 1 - \delta_{k,0} + \delta_{k,0}\delta_{s,\mathrm{sgn}(q_f B)}$ with $s = \pm 1$. Then, we can solve the phase transition from the thermodynamical potential, and the results are shown in Figure 2a, which shows that the phase transition of the chirality imbalance is of first order with $r_A = -0.3, -0.5, -0.8$ at $B = 0$. The FOPT will induce GWs and generate PBHs. The bounce solution and bubble nucleation temperature of this FOPT are solved to obtain the strength parameter α and the inverse duration time β/H, which are essential to calculate the gravitational wave power spectra in [67]. The results are shown in Table 1.

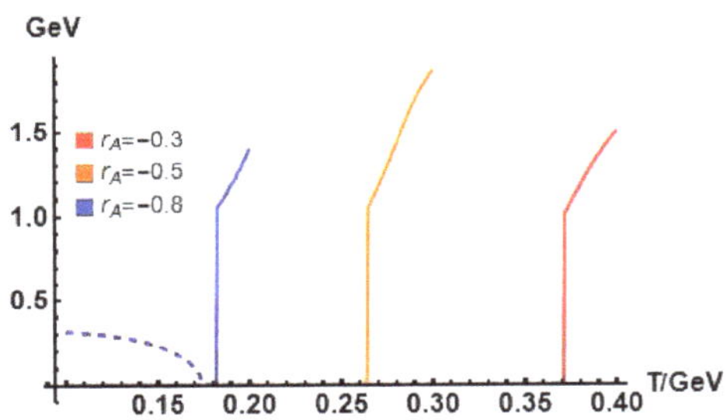

(**a**) $\mu 5$ (lines) and σ (dashed lines) as a function of T at $eB = 0$.

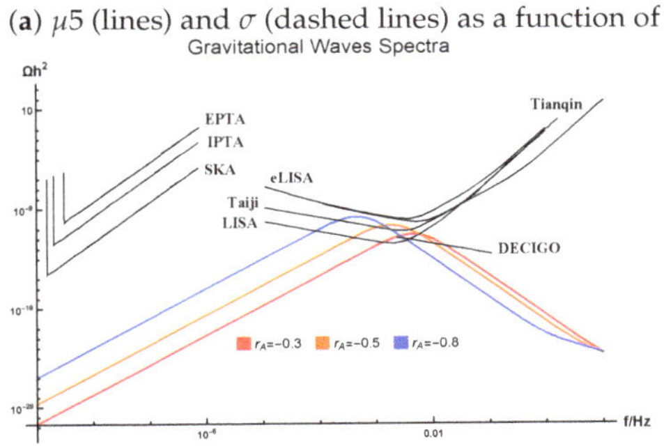

(**b**) Gravitational wave spectra.

Figure 2. (**a**) The chiral condensate σ (dashed lines) and the dynamically induced chiral chemical potential μ_5 (lines) as a function of the temperature T at $eB = 0$ for $r_A = G_A/G_S = -0.3, -0.5$ and $-0.8\,\mathrm{GeV}^2$; here, σ and μ_5 are in units of GeV. (**b**) GW spectra for $r_A = G_A/G_S = -0.3, -0.5$ and $-0.8\,\mathrm{GeV}^2$ at $eB = 0$. The figures were updated by Jingdong Shao based on the results in [67].

Table 1. Bounce solution for the nucleation temperature T_n, parameters α and $\beta/H*$ corresponding to different values of r_A and eB.

$r_A = G_A/G_S$	T_n/GeV	α	β/H_*
−0.3	0.3648	0.7343	27582
−0.5	0.2561	1.741	16274
−0.8	0.1679	4.850	6105.7

It is noticed that the value of $\beta/H*{\sim}10^4$ solved in this model is quite different from the typical value from that obtained in the EWPT, where the typical value of $\beta/H* = 1{\sim}100$ [83–85], but the holographic QCD calculations also gave $\beta/H*{\sim}10^4$ [62,64]. The large $\beta/H*$ indicates that the PT completes in an extremely short time period; thus, it is not favorable for the formation of the primordial black hole, as also proven in [67]. Figure 2b shows the GW spectra for $r_A = G_A/G_S = -0.3, -0.5$ and $-0.8\,\text{GeV}^2$ at $eB = 0$. The GW spectra from the FOPT of the chirality imbalance can be detected by Taiji, Lisa, and DECIGO.

There are also many other models: for example, the bag model has been investigated in the early universe [86,87]. For the QCD epoch in the early universe after the EWPT in the temperature region of 100 MeV∼100 GeV, we need to consider the interplay between the quark dynamics including both light flavors and heavy flavors and gluodynamics; for that purpose, the holographic QCD method might be one of the powerful theoretical tools [88–91].

3. Dense QCD Matter

Exploring the QCD phase structure and equation of state (EoS) at high baryon density have been boosted both by neutron star (NS) and heavy-ion collisions.

The search for the QCD CEP at finite baryon density has been one of the most-important goals for the BES program for RHICs, as well as for the future accelerator facilities at the FAIR and NICA. The higher-order fluctuations of conserved charges, especially the baryon number fluctuations, carry the divergence feature of the correlation length at the CEP [92,93]. The measurement of the kurtosis of the baryon number fluctuations $\kappa\sigma^2$ or C_4/C_2 from BES-I for RHICs in the collision energy $\sqrt{s_{NN}} = 200{\sim}7.7$ GeV shows a non-monotonic energy-dependent behavior [94–96]. It has been observed that $\kappa\sigma^2$ of the net proton number distributions starts from about 1 at $\sqrt{s_{NN}} = 200$ GeV, corresponding to the baryon chemical potential $\mu_B{\sim}0.1$ GeV, decreases to around 0.1 at $\sqrt{s_{NN}} = 20$ GeV, and rises quickly up to 3.5 at $\sqrt{s_{NN}} = 7.7$ GeV, corresponding to the baryon chemical potential $\mu_B{\sim}0.4$ GeV.

The three-flavor realistic PNJL (rPNJL) model, which takes into account eight-quark interaction [97], can give a realistic critical temperature $T_c = 160\,\text{MeV}$ at $\mu = 0$, and the effective potential takes the form of

$$\begin{aligned}\Omega \;=\;& g_S\sum_f \sigma_f^2 - \frac{g_D}{2}\sigma_u\sigma_d\sigma_s + 3\frac{g_1}{2}(\sum_f \sigma_f^2)^2 + 3g_2\sum_f \sigma_f^4 - 6\int_0^\Lambda \frac{d^3p}{(2\pi)^3}E_f \\ & -2T\int \frac{d^3p}{(2\pi)^3}\ln[1+3(\Phi+\bar{\Phi}e^{-(E_f-\mu_f)/T})e^{-(E_f-\mu_f)/T}+e^{-3(E_f-\mu_f)/T}] \\ & -2T\int \frac{d^3p}{(2\pi)^3}\ln[1+3(\Phi+\bar{\Phi}e^{-(E_f+\mu_f)/T})e^{-(E_f+\mu_f)/T}+e^{-3(E_f+\mu_f)/T}] \\ & +U'(\Phi,\bar{\Phi},T). \end{aligned} \tag{2}$$

where $f = u, d, s$ refers to three-flavor light quarks, $\sigma_f = \left\langle \bar{\psi}_f\psi_f \right\rangle$ is the quark–antiquark condensate for different flavors, and $E_f = \sqrt{p^2 + M_f^2}$ with M_f the constituent quark mass taking the form of $M_f = m_f - 2g_S\sigma_f + \frac{g_D}{4}\sigma_{f+1}\sigma_{f+2} - 2g_1\sigma_f(\sum_{f'}\sigma_{f'}^2) - 4g_2\sigma_f^3$. U' takes the form of $\frac{U'}{T^4} = \frac{U}{T^4} - \kappa\ln[J(\Phi,\bar{\Phi})]$ and its detailed description can be found in [98,99].

The three-flavor rPNJL model [97] predicts that the CEP of the chiral phase transition is located at $(\mu_B^E = 720\text{ MeV}, T^E = 93\text{ MeV})$ [99,100]. The chiral phase transition boundary in this rPNJL model is close to the extracted freeze-out line from heavy-ion collisions [101,102], and the kurtosis of baryon number fluctuations $\kappa\sigma^2$ calculated in the rPNJL model along the experimental freeze-out line [101,102] agrees well with the BES-I measurement and predicts a peak structure around $\sqrt{s_{NN}} = 5$ GeV; see Figure 3. It was emphasized in [99,100] that the dip structure of $\kappa\sigma^2$ can be explained at the low-baryon-density region, the freeze-out line lies above the chiral phase transition line and crosses the chiral phase transition line, while the peak of $\kappa\sigma^2$ is a clean signature for the existence of the CEP.

Recently, the results at $\sqrt{s_{NN}} = 3$ GeV and the measurement of HADES at $\sqrt{s_{NN}} = 2.4$ GeV were updated in [103]. The data indicate a peak structure $\sqrt{s_{NN}} = 5$ GeV, and more data with higher precision are expected at lower collision energies. It is noted that the data at $\sqrt{s_{NN}} = 3$ GeV have the tendency to be negative, which might indicate another CEP of the liquid–gas phase transition.

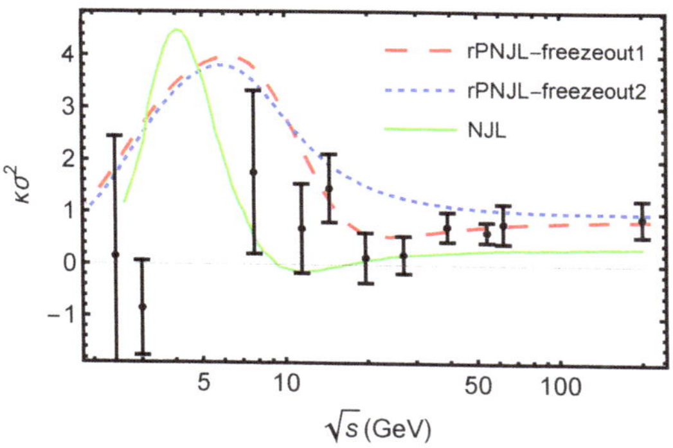

Figure 3. The $\kappa\sigma^2$ produced in the rPNJL model [99,100], as well as in the NJL model [104] as a function of $\sqrt{s}$ along the freeze-out lines, compared with the updated measurement of C4/C2 or $\kappa\sigma^2$ for the proton and net-proton at the RHIC BES-II program in [103]. The figure was updated by Kun Xu based on the results in [99,100].

Dense QCD matter may exist inside the neutron star (NS), which is the remnant after a massive super-giant star collapses. In 1967, the discovery of radio pulsars confirmed the existence of NSs. Since then, more than 2700 radio pulsars have been detected. From the experimental side, the mass and radius of the NSs can be measured with high precision [105–115]. Recent developments in observational astronomy including the operating satellites, the Neutron star Interior Composition Explorer (NICER), and along with the observation of the GWs, the Advanced LIGO, Virgo, and KAGRA [116,117] have also provided high precision measurements on neutron star masses and radii from the tidal deformability measurements. These high precision measurement gives very strict constraints on the equation of state (EoS) of dense QCD matter inside neutron stars.

The baryon number density n_B in the core of the NS can reach 5$\sim$10 times of the nuclear saturation density n_s, which lies in nonperturbative QCD region. From theory side, it is quite challenging to explore the internal structure of NSs. There have been many efforts trying to describe the EoS of cold QCD dense matter including nuclear matter and quark matter by using effective models. In recent years, holographic QCD method has been applied in describing nonperturbative QCD physics. In Ref. [118], the hybrid EoS of the cold strong interaction matter has been constructed using holographic models. In the low baryon number density n_B range, Hebeler-Lattimer-Pethick-Schwenk (HLPS) can be used to describe the nuclear matter EoS. At the intermediate n_B region, the holographic QCD is used to calculate the EoS. Our holographic QCD model consists of the Einstein-Maxwell-dilaton (EMD) system for the gluodynamical background and the improved Karch-Katz-Son-Stephanov (KKSS) action for the flavor part to describe the holographic nuclear matter and the holographic quark matter. In the framework of the Einstein-Maxwell-dilaton (EMD) system, i.e., the gravity-dilaton coupling system at zero chemical potential extended to finite chemical potential. The total action of the 5-dimensional holographic QCD model is

$$S^s_{\text{total}} = S^s_{EMD} + S^s_f, \tag{3}$$

where S^s_{EMD} is the action for the EMD system in the string frame, and S^s_f is the action that describes the flavor part in the string frame. In order to investigate the equation of state, it is more convenient to work in the Einstein frame. By using the relation $g^s_{MN} = e^{\frac{4}{3}\Phi} g^E_{MN}$, the EMD background action in the Einstein frame has the form of

$$S^E_{EMD} = \frac{1}{2\kappa_5^2} \int d^5x \sqrt{-g^E} \left[R^E - \frac{4}{3} g^{EMN} \partial_M \Phi \partial_N \Phi - V^E(\Phi) - \frac{h(\Phi)}{4} g^{EM\tilde{M}} g^{EN\tilde{N}} F_{MN} F_{\tilde{M}\tilde{N}} \right], \tag{4}$$

with $V^E = e^{\frac{4}{3}\Phi} V^s$. The flavor part or the matter part is given by the meson part and baryon part $S^s_f = S^s_{\text{mesons}} + S^s_{\text{baryons}}$. It is expected that S^s_{baryons} is small, and the meson part can be described by the KKSS model [119]:

$$S^s_f = -\beta \int d^5x \sqrt{-g^s} e^{-\Phi} \left\{ |D_M X|^2 + V^s_X(X_M) + \frac{1}{4g_5^2} \left(F_L^2 + F_R^2 \right) \right\}, \tag{5}$$

where X is the 5-dimensional scalar field. In the holographic model, we take the dilation field $\phi(z) = c_1 z^2$, and the function $h_\phi(z) = e^{c_2 z^2 - A_E(z)}$, with $c_1 = 1.536\,\text{GeV}^2$ and $c_2 = 1.16\,\text{GeV}^2$. The 5-dimensional Newtonian constant is taken as $G_5 = 17.930$.

The hybrid model gives the mass-radius relation and the tidal deformability of the neutron stars which are in agreement with astrophysical measurements as shown in Figure 4a,b. The possible maximum mass for the neutron star is about $2.5 M_\odot$ and the radius is about 12 km. In the current holographic model, we didn't consider the phase transition between the nuclear matter and the quark matter. The holographic nuclear matter here might refer to the quarkyonic phase [120,121], which is in confinement with chiral symmetry restoration.

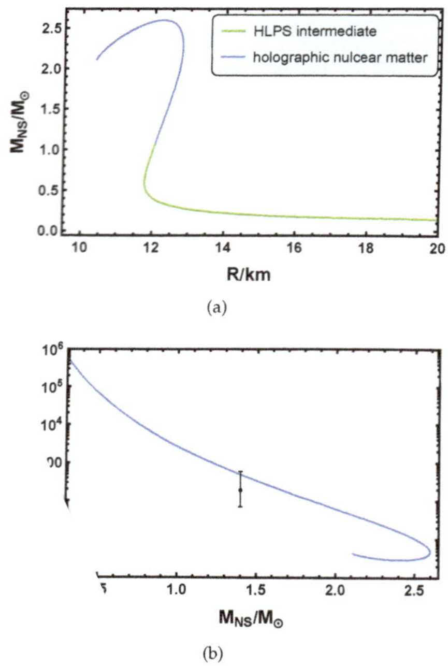

ıss-radius relation in (**a**) and tidal deformability as a function of the mass ratio ı star and the sun in (**b**) calculated in a dynamical holographic QCD model. The ıy Lin Zhang based on results in [118].

4. QCD Matter under Electromagnetic and Vortical Field

The hot and dense matter created in non-central heavy-ion collisions (HICs) shows the fastest rotation with the angular momentum in the range of $\mathcal{O}(10^4)\hbar$–$\mathcal{O}(10^5)\hbar$ or a vorticity of order $10^{21}\,\mathrm{s}^{-1}$[33,122–124], and the strongest magnetic field $B{\sim}10^{18}G$ for RHICs, while $B{\sim}10^{20}G$ at the LHC [30,31]. Both the magnetic field and rotation will polarize quarks and gluons, thus changing the QCD matter properties under an external magnetic field and rotation.

The novel phenomena of QCD matter under external magnetic fields include the chiral magnetic effect (CME) [75–77], a magnetic field enhancing the chiral symmetry breaking in the vacuum, which is called magnetic catalysis (MC) [125–127], and the critical temperature of the chiral phase transition decreasing with external magnetic fields, i.e., inverse magnetic catalysis (IMC) [128–130]. Recent lattice calculations showed some unexpected properties of pion mass under a magnetic field: (1) the neutral pion mass decreases with the magnetic field, then saturates with a strong magnetic field; (2) the charged pion mass exhibits a nonmonotonic magnetic-field-dependent behavior [131]; it first increases and reaches a peak around $eB{\sim}0.5\,\mathrm{GeV}^2$, then decreases. Besides, the lattice calculation also showed that the magnetic susceptibility of quark matter is negative (diamagnetism) at a low temperature and is positive (paramagnetism) at a high temperature [132,133]. Much effort has been made to understand these properties, and it is still challenging to describe all these phenomena within one self-consistent framework.

The global polarization of Λ and $\bar{\Lambda}$ was predicted in [34] and was observed by the STAR Collaboration for Au+Au collisions at $\sqrt{s_{NN}} = 7.7$–$200\,\mathrm{GeV}$ [35]. The experimental data showed that: (1) both the global spin polarization and the splitting of $P_{\Lambda/\bar{\Lambda}}$ become stronger at lower collision energy; (2) the polarization of $\bar{\Lambda}$ is larger than that of Λ, i.e., $P_{\bar{\Lambda}} > P_{\Lambda}$.

In [134], it was found that the global polarization of $\Lambda/\bar{\Lambda}$ can be induced by the rotation and the splitting of the $\Lambda - \bar{\Lambda}$ polarization induced by a magnetic field. In [134], the three-flavor NJL model included the 't Hooft determinant term with the Lagrangian $\mathcal{L}_\chi$:

$$\mathcal{L}_\chi = \frac{G_S}{2}\sum_{a=0}^{8}\Big\{(\bar{\psi}\lambda^a\psi)^2 + (\bar{\psi}i\gamma^5\lambda^a\psi)^2\Big\} - G_K\Big\{\det\bar{\psi}(1+\gamma^5)\psi + \det\bar{\psi}(1-\gamma^5)\psi\Big\}, \tag{6}$$

taking into account the anomalous magnetic moment (AMM) of quarks under an external magnetic field B:

$$\mathcal{L}_{AMM} = \bar{\psi}\Big(i\gamma^\mu D_\mu - \hat{m} + \kappa q B\sigma^{12}\Big)\psi, \tag{7}$$

and considering the contribution of the axial vector interaction under rotation characterized by the angular velocity Ω:

$$\mathcal{L}_A = \bar{\psi}\Big[\gamma^0(i\partial_t + \Omega\hat{J}_z) + \mu\gamma^0\Big]\psi - G_A\Big\{(\bar{\psi}\gamma^\mu\psi)^2 + (\bar{\psi}\gamma^\mu\gamma^5\psi)^2\Big\}. \tag{8}$$

With the values of Ω and eB extracted from experimental measurement, we input these values into the above dynamical quark model including the axial vector interaction and AMM and re-derived the polarization of the s quark and $\bar{s}$ quark, and the results are shown in Figure 5. It was noticed that the rotation contributes the same polarization to the s and $\bar{s}$ quarks, while the magnetic field contributes a positive polarization for the $\bar{s}$ quark, while giving a negative polarization for the s quark. The splitting of the $\Lambda - \bar{\Lambda}$ polarization induced by the magnetic field includes two sources: (1) the remaining magnetic field from the evolution of the system after the collision and (2) the magnetic field induced by the vorticity [135]; this is shown in Figure 6.

However, at a collision energy below 7.7 GeV, the magnitude of the magnetic field used at freeze-out is at least 20-times smaller to induce the large splitting polarization of $P_{\Lambda/\bar{\Lambda}}$. It will be needed in the future to reanalyze the magnetic field evolution at low collision energies where the vorticity is high; thus, the generation of the magnetic field from the vorticity [135] should be taken into account simultaneously.

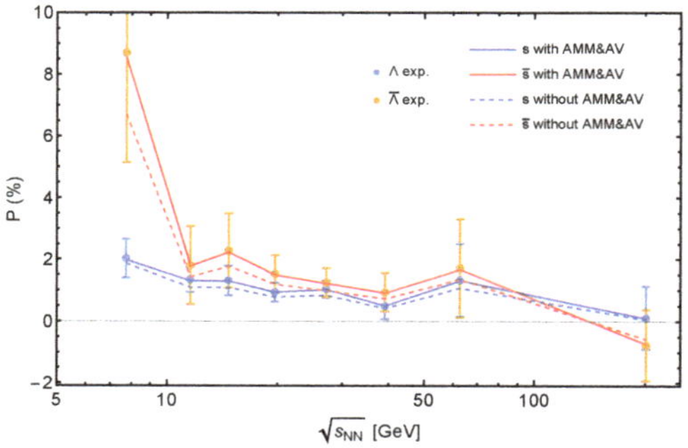

Figure 5. The spin polarization of the s quark and $\bar{s}$ quark induced by rotation and a magnetic field as a function of the collision energy $\sqrt{s}$. The figure is from [134], where the details can be found.

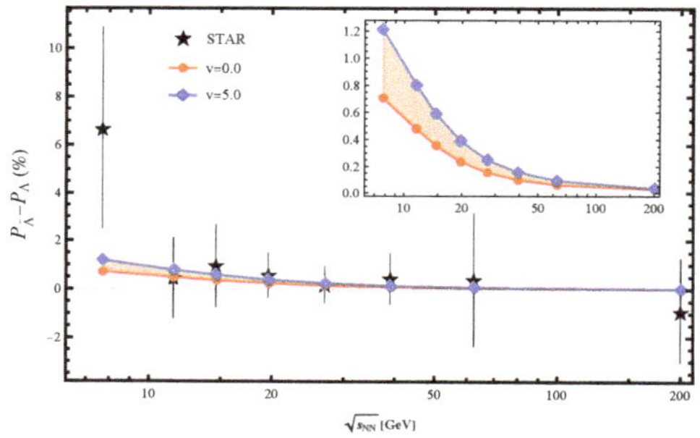

Figure 6. The splitting of $\Lambda/\bar{\Lambda}$ polarization $P_{\bar{\Lambda}} - P_{\Lambda}$ induced by rotation and a magnetic field as a function of the collision energy $\sqrt{s}$. The figure is from [134], where the details can be found.

5. Outlook

As we have shown, understanding quark matter's properties and QCD phase transitions under extreme conditions is of special interest for investigating the evolution of the universe, exploring the structure of compact stars, and understanding observable heavy-ion collisions. The properties of QCD matter and phase transitions in which we are interested lie in the nonperturbative QCD region, where the interplay of the chiral dynamics and gluodynamics plays an essential role. The lattice QCD calculation can provide first-principles results, though there are still challenges in the finite baryon density region. Four-dimensional effective QCD models, especially chiral effective models, can provide some qualitative analysis of or insight into possible interesting phenomena. Five-dimensional holographic QCD models based on the AdS/CFT correspondence have the potential to unveil the interplay of chiral dynamics and gluodynamics.

Author Contributions: These authors contributed equally to this work. All authors have read and agreed to the published version of the manuscript.

Funding: We thank Jingdong Shao, Kun Xu and LIn Zhang for the help to update the figures. This work is supported in part by NSFC Grant Nos. 12235016, 12221005, 11725523, 11735007, 11890712 and the Strategic Priority Research Program of Chinese Academy of Sciences under Grant Nos XDB34030000, the start-up funding from University of Chinese Academy of Sciences(UCAS), and the Fundamental Research Funds for the Central Universities.

Acknowledgments: We thank Jingdong Shao, Kun Xu, and Lin Zhang for the help in updating the figures.

Conflicts of Interest: The authors declare no conflict of interest.

References

1. Gross, D.J.; Wilczek, F. Ultraviolet Behavior of Nonabelian Gauge Theories. *Phys. Rev. Lett.* **1973**, *30*, 1343–1346. [CrossRef]
2. Politzer, H.D. Reliable Perturbative Results for Strong Interactions? *Phys. Rev. Lett.* **1973**, *30*, 1346–1349. [CrossRef]
3. Schwarz, D.J. The first second of the universe. *Ann. Phys.* **2003**, *12*, 220–270. [CrossRef]
4. Baym, G.; Chin, S.A. Can a Neutron Star Be a Giant MIT Bag? *Phys. Lett. B* **1976**, *62*, 241–244. [CrossRef]
5. Collins, J.C.; Perry, M.J. Superdense Matter: Neutrons Or Asymptotically Free Quarks? *Phys. Rev. Lett.* **1975**, *34*, 1353. [CrossRef]
6. Luo, X.; Wang, Q.; Xu, N.; Zhuang, P. (Eds.) *Properties of QCD Matter at High Baryon Density*; Springer: Berlin/Heidelberg, Germany, 2022. [CrossRef]
7. Karsch, F. Lattice QCD at Non-Zero Temperature and Density. 2022. Available online: http://xxx.lanl.gov/abs/2212.03015 (accessed on 30 December 2022).
8. Aarts, G.; Aichelin, J.; Allton, C.; Athenodorou, A.; Bachtis, D.; Bonanno, C.; Brambilla, N.; Bratkovskaya, E.; Bruno, M.; Caselle, M.; et al. Phase Transitions in Particle Physics–Results and Perspectives from Lattice Quantum Chromo-Dynamics. *arXiv* **2023**, arXiv:2301.04382.
9. de Forcrand, P.; D'Elia, M. Continuum limit and universality of the Columbia plot. *PoS* **2017**, *LATTICE2016*, 081. [CrossRef]
10. Karsch, F. Lattice QCD at high temperature and density. *Lect. Notes Phys.* **2002**, *583*, 209–249. [CrossRef]
11. Aoki, Y.; Endrodi, G.; Fodor, Z.; Katz, S.D.; Szabo, K.K. The Order of the quantum chromodynamics transition predicted by the standard model of particle physics. *Nature* **2006**, *443*, 675–678. [CrossRef]
12. Allton, C.R.; Ejiri, S.; Hands, S.J.; Kaczmarek, O.; Karsch, F.; Laermann, E.; Schmidt, C.; Scorzato, L. The QCD thermal phase transition in the presence of a small chemical potential. *Phys. Rev. D* **2002**, *66*, 074507, [CrossRef]
13. Allton, C.R.; Doring, M.; Ejiri, S.; Hands, S.J.; Kaczmarek, O.; Karsch, F.; Laermann, E.; Redlich, K. Thermodynamics of two flavor QCD to sixth order in quark chemical potential. *Phys. Rev. D* **2005**, *71*, 054508. [CrossRef]
14. Borsanyi, S.; Endrodi, G.; Fodor, Z.; Katz, S.D.; Krieg, S.; Ratti, C.; Szabo, K.K. QCD equation of state at nonzero chemical potential: Continuum results with physical quark masses at order mu^2. *J. High Energy Phys.* **2012**, *08*, 053. [CrossRef]
15. Bazavov, A.; Ding, H.T.; Hegde, P.; Kaczmarek, O.; Karsch, F.; Laermann, E.; Maezawa, Y.; Mukherjee, S.; Ohno, H.; Petreczky, P.; et al. The QCD Equation of State to $\mathcal{O}(\mu_B^6)$ from Lattice QCD. *Phys. Rev. D* **2017**, *95*, 054504. [CrossRef]
16. Borsányi, S.; Fodor, Z.; Guenther, J.N.; Kara, R.; Katz, S.D.; Parotto, P.; Pásztor, A.; Ratti, C.; Szabó, K.K. Lattice QCD equation of state at finite chemical potential from an alternative expansion scheme. *Phys. Rev. Lett.* **2021**, *126*, 232001. [CrossRef]
17. Pisarski, R.D.; Wilczek, F. Remarks on the Chiral Phase Transition in Chromodynamics. *Phys. Rev. D* **1984**, *29*, 338–341. [CrossRef]
18. Hatta, Y.; Ikeda, T. Universality, the QCD critical/tricritical point and the quark number susceptibility. *Phys. Rev. D* **2003**, *67*, 014028. [CrossRef]
19. Klevansky, S.P. The Nambu-Jona-Lasinio model of quantum chromodynamics. *Rev. Mod. Phys.* **1992**, *64*, 649–708. [CrossRef]
20. Zhuang, P.; Huang, M.; Yang, Z. Density effect on hadronization of a quark plasma. *Phys. Rev. C* **2000**, *62*, 054901. [CrossRef]
21. Fukushima, K. Phase diagrams in the three-flavor Nambu-Jona-Lasinio model with the Polyakov-loop. *Phys. Rev. D* **2008**, *77*, 114028; Erratum in *Phys. Rev. D* **2008**, *78*, 039902. [CrossRef]
22. Ratti, C.; Roessner, S.; Thaler, M.A.; Weise, W. Thermodynamics of the PNJL model. *Eur. Phys. J. C* **2007**, *49*, 213–217. [CrossRef]
23. Shao, G.; Tang, Z.; Gao, X.; He, W. Baryon number fluctuations and the phase structure in the PNJL model. *Eur. Phys. J. C* **2018**, *78*, 138. [CrossRef]
24. Tawfik, A. QCD phase diagram: A Comparison of lattice and hadron resonance gas model calculations. *Phys. Rev. D* **2005**, *71*, 054502. [CrossRef]
25. Redlich, K.; Karsch, F.; Tawfik, A. Heavy ion collisions and lattice QCD at finite baryon density. *J. Phys. G* **2004**, *30*, S1271–S1274. [CrossRef]
26. Tawfik, A. The Influence of strange quarks on QCD phase diagram and chemical freeze-out: Results from the hadron resonance gas model. *J. Phys. G* **2005**, *31*, S1105–S1110. [CrossRef]
27. Xin, X.y.; Qin, S.x.; Liu, Y.x. Quark number fluctuations at finite temperature and finite chemical potential via the Dyson–Schwinger equation approach. *Phys. Rev. D* **2014**, *90*, 076006. [CrossRef]
28. Fu, W.j.; Wu, Y.l. Fluctuations and Correlations of Conserved Charges near the QCD Critical Point. *Phys. Rev. D* **2010**, *82*, 074013. [CrossRef]
29. Li, Z.; Chen, Y.; Li, D.; Huang, M. Locating the QCD critical end point through the peaked baryon number susceptibilities along the freeze-out line. *Chin. Phys. C* **2018**, *42*, 013103. [CrossRef]
30. Skokov, V.; Illarionov, A.Y.; Toneev, V. Estimate of the magnetic field strength in heavy-ion collisions. *Int. J. Mod. Phys. A* **2009**, *24*, 5925–5932. [CrossRef]
31. Deng, W.T.; Huang, X.G. Event-by-event generation of electromagnetic fields in heavy-ion collisions. *Phys. Rev. C* **2012**, *85*, 044907. [CrossRef]
32. Chen, H.L.; Huang, X.G.; Liao, J. QCD phase structure under rotation. *Lect. Notes Phys.* **2021**, *987*, 349–379. [CrossRef]

33. Becattini, F.; Piccinini, F.; Rizzo, J. Angular momentum conservation in heavy-ion collisions at very high energy. *Phys. Rev. C* **2008**, *77*, 024906. [CrossRef]
34. Liang, Z.T.; Wang, X.N. Globally polarized quark-gluon plasma in non-central A+A collisions. *Phys. Rev. Lett.* **2005**, *94*, 102301; Erratum in *Phys. Rev. Lett.* **2006**, *96*, 039901. [CrossRef] [PubMed]
35. Adamczyk, L.; The STAR Collaboration. Global Λ hyperon polarization in nuclear collisions: Evidence for the most vortical fluid. *Nature* **2017**, *548*, 62–65. [CrossRef]
36. Abbott, B.P.; Abbott, R.; Abbott, T.D.; Abernathy, M.R.; Acernese, F.; Ackley, K.; Adams, C.; Adams, T.; Addesso, P.; Adhikari, R.X.; et al. Observation of Gravitational Waves from a Binary Black Hole Merger. *Phys. Rev. Lett.* **2016**, *116*, 061102. [CrossRef] [PubMed]
37. Einstein, A. Approximative Integration of the Field Equations of Gravitation. *Sitzungsber. Preuss. Akad. Wiss. Berlin (Math. Phys.)* **1916**, *1916*, 688–696.
38. Einstein, A. Über Gravitationswellen. *Sitzungsber. Preuss. Akad. Wiss. Berlin (Math. Phys.)* **1918**, *1918*, 154–167.
39. Rummukainen, K. Finite T electroweak phase transition on the lattice. *Nucl. Phys. B Proc. Suppl.* **1997**, *53*, 30–42. [CrossRef]
40. Vaskonen, V. Electroweak baryogenesis and gravitational waves from a real scalar singlet. *Phys. Rev. D* **2017**, *95*, 123515. [CrossRef]
41. Beniwal, A.; Lewicki, M.; Wells, J.D.; White, M.; Williams, A.G. Gravitational wave, collider and dark matter signals from a scalar singlet electroweak baryogenesis. *J. High Energy Phys.* **2017**, *08*, 108. [CrossRef]
42. Beniwal, A.; Lewicki, M.; White, M.; Williams, A.G. Gravitational waves and electroweak baryogenesis in a global study of the extended scalar singlet model. *J. High Energy Phys.* **2019**, *02*, 183. [CrossRef]
43. Alves, A.; Ghosh, T.; Guo, H.K.; Sinha, K.; Vagie, D. Collider and Gravitational Wave Complementarity in Exploring the Singlet Extension of the Standard Model. *J. High Energy Phys.* **2019**, *04*, 052. [CrossRef]
44. Cline, J.M.; Lemieux, P.A. Electroweak phase transition in two Higgs doublet models. *Phys. Rev.* **1997**, *D55*, 3873–3881. [CrossRef]
45. Basler, P.; Krause, M.; Muhlleitner, M.; Wittbrodt, J.; Wlotzka, A. Strong First Order Electroweak Phase Transition in the CP-Conserving 2HDM Revisited. *J. High Energy Phys.* **2017**, *02*, 121. [CrossRef]
46. Dorsch, G.C.; Huber, S.J.; Konstandin, T.; No, J.M. A Second Higgs Doublet in the Early Universe: Baryogenesis and Gravitational Waves. *J. Cosmol. Astropart. Phys.* **2017**, *1705*, 052. [CrossRef]
47. Tawfik, A.N.; Greiner, C. Early Universe Thermodynamics and Evolution in Nonviscous and Viscous Strong and Electroweak epochs: Possible Analytical Solutions. *Entropy* **2021**, *23*, 295. [CrossRef]
48. Tawfik, A.; Harko, T. Quark-Hadron Phase Transitions in Viscous Early Universe. *Phys. Rev. D* **2012**, *85*, 084032. [CrossRef]
49. Tawfik, A. Cosmological Consequences of QCD Phase Transition(s) in Early Universe. *AIP Conf. Proc.* **2009**, *1115*, 239–247. [CrossRef]
50. Tawfik, A.; Wahba, M.; Mansour, H.; Harko, T. Viscous Quark-Gluon Plasma in the Early Universe. *Annalen Phys.* **2011**, *523*, 194–207. [CrossRef]
51. Hogan, C.J. Gravitational radiation from cosmological phase transitions. *Mon. Not. R. Astron. Soc.* **1986**, *218*, 629–636. [CrossRef]
52. Kosowsky, A.; Turner, M.S.; Watkins, R. Gravitational waves from first order cosmological phase transitions. *Phys. Rev. Lett.* **1992**, *69*, 2026–2029. [CrossRef]
53. Kosowsky, A.; Turner, M.S. Gravitational radiation from colliding vacuum bubbles: Envelope approximation to many bubble collisions. *Phys. Rev. D* **1993**, *47*, 4372–4391. [CrossRef] [PubMed]
54. Kamionkowski, M.; Kosowsky, A.; Turner, M.S. Gravitational radiation from first order phase transitions. *Phys. Rev. D* **1994**, *49*, 2837–2851. [CrossRef] [PubMed]
55. Hawking, S.W.; Moss, I.G.; Stewart, J.M. Bubble Collisions in the Very Early Universe. *Phys. Rev. D* **1982**, *26*, 2681. [CrossRef]
56. Jung, T.H.; Okui, T. Primordial black holes from bubble collisions during a first-order phase transition. *arXiv* **2021**, arXiv:2110.04271.
57. Khlopov, M.Y.; Konoplich, R.; Rubin, S.; Sakharov, A.S. First order phase transitions as a source of black holes in the early universe. *arXiv* **1999**, arXiv:hep-ph/9912422.
58. Carr, B.; Clesse, S.; García-Bellido, J. Primordial black holes from the QCD epoch: Linking dark matter, baryogenesis and anthropic selection. *Mon. Not. R. Astron. Soc.* **2021**, *501*, 1426–1439. [CrossRef]
59. Byrnes, C.T.; Hindmarsh, M.; Young, S.; Hawkins, M.R.S. Primordial black holes with an accurate QCD equation of state. *J. Cosmol. Astropart. Phys.* **2018**, *08*, 041. [CrossRef]
60. Jedamzik, K. Primordial black hole formation during the QCD epoch. *Phys. Rev. D* **1997**, *55*, 5871–5875. [CrossRef] [CrossRef]
61. Profumo, S.; Ramsey-Musolf, M.J.; Shaughnessy, G. Singlet Higgs phenomenology and the electroweak phase transition. *J. High Energy Phys.* **2007**, *8*, 010. [CrossRef]
62. Chen, Y.; Huang, M.; Yan, Q.S. Gravitation waves from QCD and electroweak phase transitions. *J. High Energy Phys.* **2018**, *5*, 178. [CrossRef]
63. He, S.; Li, L.; Li, Z.; Wang, S.J. Gravitational Waves and Primordial Black Hole Productions from Gluodynamics. *arXiv* **2022**, arXiv:2210.14094.
64. Morgante, E.; Ramberg, N.; Schwaller, P. Echo of the Dark: Gravitational Waves from Dark SU(3) Yang-Mills Theory. *arXiv* **2022**, arXiv:2210.11821.

65. Asadi, P.; Kramer, E.D.; Kuflik, E.; Slatyer, T.R.; Smirnov, J. Glueballs in a thermal squeezeout model. *J. High Energy Phys.* **2022**, *7*, 006. [CrossRef]
66. Carenza, P.; Pasechnik, R.; Salinas, G.; Wang, Z.W. Glueball Dark Matter revisited. *Phys. Rev. Lett.* **2022**, *129*, 26. [CrossRef] [PubMed]
67. Shao, J.; Huang, M. Gravitational waves and primordial black holes from chirality imbalanced QCD first-order phase transition with $\mathcal{P}$ and $\mathcal{CP}$ violation. *Phys. Rev. D* **2023**, *107*, 043011. [CrossRef]
68. 't Hooft, G. Symmetry Breaking Through Bell-Jackiw Anomalies. *Phys. Rev. Lett.* **1976**, *37*, 8–11. . [CrossRef]
69. 't Hooft, G. How Instantons Solve the U(1) Problem. *Phys. Rep.* **1986**, *142*, 357–387. [CrossRef]
70. Belavin, A.A.; Polyakov, A.M.; Schwartz, A.S.; Tyupkin, Y.S. Pseudoparticle Solutions of the Yang-Mills Equations. *Phys. Lett. B* **1975**, *59*, 85–87. [CrossRef]
71. Witten, E. Current Algebra Theorems for the U(1) Goldstone Boson. *Nucl. Phys. B* **1979**, *156*, 269–283. [CrossRef]
72. Veneziano, G. U(1) Without Instantons. *Nucl. Phys. B* **1979**, *159*, 213–224. [CrossRef]
73. Vicari, E.; Panagopoulos, H. Theta dependence of SU(N) gauge theories in the presence of a topological term. *Phys. Rept.* **2009**, *470*, 93–150. [CrossRef]
74. Schäfer, T.; Shuryak, E.V. Instantons in QCD. *Rev. Mod. Phys.* **1998**, *70*, 323–426. [CrossRef]
75. Kharzeev, D.; Zhitnitsky, A. Charge separation induced by P-odd bubbles in QCD matter. *Nucl. Phys. A* **2007**, *797*, 67–79. [CrossRef]
76. Kharzeev, D.E.; McLerran, L.D.; Warringa, H.J. The Effects of topological charge change in heavy-ion collisions: 'Event by event P and CP violation'. *Nucl. Phys. A* **2008**, *803*, 227–253. [CrossRef]
77. Fukushima, K.; Kharzeev, D.E.; Warringa, H.J. The Chiral Magnetic Effect. *Phys. Rev. D* **2008**, *78*, 074033. [CrossRef]
78. Ilgenfritz, E.M.; Shuryak, E.V. Chiral Symmetry Restoration at Finite Temperature in the Instanton Liquid. *Nucl. Phys. B* **1989**, *319*, 511–520. [CrossRef]
79. Ilgenfritz, E.M.; Shuryak, E.V. Quark induced correlations between instantons drive the chiral phase transition. *Phys. Lett. B* **1994**, *325*, 263–266. [CrossRef]
80. Schäfer, T.; Shuryak, E.V.; Verbaarschot, J.J.M. The Chiral phase transition and instanton—Anti-instanton molecules. *Phys. Rev. D* **1995**, *51*, 1267–1281. [CrossRef]
81. Zhang, Z. Correction to the Chiral Magnetic Effect from axial vector interaction. *Phys. Rev. D* **2012**, *85*, 114028. [CrossRef]
82. Yu, L.; Liu, H.; Huang, M. Spontaneous generation of local CP violation and inverse magnetic catalysis. *Phys. Rev. D* **2014**, *90*, 074009. [CrossRef]
83. Ellis, J.; Lewicki, M.; No, J.M. On the Maximal Strength of a First-Order Electroweak Phase Transition and its Gravitational Wave Signal. *J. Cosmol. Astropart. Phys.* **2019**, *04*, 003. [CrossRef]
84. Leitao, L.; Megevand, A. Gravitational waves from a very strong electroweak phase transition. *J. Cosmol. Astropart. Phys.* **2016**, *05*, 037.
85. Chala, M.; Krause, C.; Nardini, G. Signals of the electroweak phase transition at colliders and gravitational wave observatories. *J. High Energy Phys.* **2018**, *07*, 062. [CrossRef]
86. DeGrand, T.A.; Kajantie, K. Supercooling, Entropy Production and Bubble Kinetics in the Quark—Hadron Phase Transition in the Early Universe. *Phys. Lett. B* **1984**, *147*, 273–278. [CrossRef]
87. Schettler, S.; Boeckel, T.; Schaffner-Bielich, J. Imprints of the QCD Phase Transition on the Spectrum of Gravitational Waves. *Phys. Rev. D* **2011**, *83*, 064030. [CrossRef]
88. Chen, Y.; Li, D.; Huang, M. The dynamical holographic QCD method for hadron physics and QCD matter. *Commun. Theor. Phys.* **2022**, *74*, 097201. [CrossRef]
89. Gursoy, U.; Kiritsis, E. Exploring improved holographic theories for QCD: Part I. *J. High Energy Phys.* **2008**, *02*, 032. [CrossRef]
90. Gursoy, U.; Kiritsis, E.; Nitti, F. Exploring improved holographic theories for QCD: Part II. *J. High Energy Phys.* **2008**, *02*, 019. [CrossRef]
91. Gubser, S.S.; Nellore, A.; Pufu, S.S.; Rocha, F.D. Thermodynamics and bulk viscosity of approximate black hole duals to finite temperature quantum chromodynamics. *Phys. Rev. Lett.* **2008**, *101*, 131601. [CrossRef]
92. Stephanov, M.A. Non-Gaussian fluctuations near the QCD critical point. *Phys. Rev. Lett.* **2009**, *102*, 032301. [CrossRef]
93. Stephanov, M.A. On the sign of kurtosis near the QCD critical point. *Phys. Rev. Lett.* **2011**, *107*, 052301. [CrossRef]
94. Adamczyk, L.; Adkins, J.K.; Agakishiev, G.; Aggarwal, M.M.; Ahammed, Z.; Alekseev, I.; Alford, J.; Anson, C.D.; Aparin, A.; Arkhipkin, D.; et al. Energy Dependence of Moments of Net-proton Multiplicity Distributions at RHIC. *Phys. Rev. Lett.* **2014**, *112*, 032302. [CrossRef] [PubMed]
95. Aggarwal, M.M.; Ahammed, Z.; Alakhverdyants, A.V.; Alekseev, I.; Alford, J.; Anderson, B.D.; Arkhipkin, D.; Averichev, G.S.; Balewski, J.; Barnby, L.S.; et al. Higher Moments of Net-proton Multiplicity Distributions at RHIC. *Phys. Rev. Lett.* **2010**, *105*, 022302. [CrossRef] [PubMed]
96. Luo, X.; Xu, N. Search for the QCD Critical Point with Fluctuations of Conserved Quantities in Relativistic Heavy-Ion Collisions at RHIC: An Overview. *Nucl. Sci. Tech.* **2017**, *28*, 112. [CrossRef]
97. Bhattacharyya, A.; Ghosh, S.K.; Maity, S.; Raha, S.; Ray, R.; Saha, K.; Upadhaya, S. Reparametrizing the Polyakov–Nambu–Jona-Lasinio model. *Phys. Rev. D* **2017**, *95*, 054005. [CrossRef]

98. Ghosh, S.K.; Mukherjee, T.K.; Mustafa, M.G.; Ray, R. PNJL model with a Van der Monde term. *Phys. Rev. D* **2008**, *77*, 094024. [CrossRef]
99. Xu, K.; Li, Z.; Huang, M. QCD critical end point from a realistic PNJL model. *EPJ Web Conf.* **2018**, *192*, 00019. [CrossRef]
100. Li, Z.; Xu, K.; Wang, X.; Huang, M. The kurtosis of net baryon number fluctuations from a realistic Polyakov–Nambu–Jona-Lasinio model along the experimental freeze-out line. *Eur. Phys. J. C* **2019**, *79*, 245. [CrossRef]
101. Das, S. Identified particle production and freeze-out properties in heavy-ion collisions at RHIC Beam Energy Scan program. *EPJ Web Conf.* **2015**, *90*, 08007. [CrossRef]
102. Begun, V.V.; Vovchenko, V.; Gorenstein, M.I. Updates to the p+p and A+A chemical freeze-out lines from the new experimental data. *J. Phys. Conf. Ser.* **2017**, *779*, 012080. [CrossRef]
103. Abdallah, M.S.; Aboona, B.E.; Adam, J.; Adamczyk, L.; Adams, J.R.; Adkins, J.K.; Agakishiev, G.; Aggarwal, I.; Aggarwal, M.M.; Ahammed, Z.; et al. Measurements of Proton High Order Cumulants in $\sqrt{s_{\rm NN}}$ = 3 GeV Au+Au Collisions and Implications for the QCD Critical Point. *Phys. Rev. Lett.* **2022**, *128*, 202303. [CrossRef]
104. Fan, W.; Luo, X.; Zong, H.S. Mapping the QCD phase diagram with susceptibilities of conserved charges within Nambu–Jona-Lasinio model. *Int. J. Mod. Phys. A* **2017**, *32*, 1750061. [CrossRef]
105. Shapiro, I.I. Fourth Test of General Relativity. *Phys. Rev. Lett.* **1964**, *13*, 789–791. [CrossRef]
106. Demorest, P.; Pennucci, T.; Ransom, S.; Roberts, M.; Hessels, J. Shapiro Delay Measurement of A Two Solar Mass Neutron Star. *Nature* **2010**, *467*, 1081–1083. [CrossRef] [PubMed]
107. Fonseca, E.; Pennucci, T.T.; Ellis, J.A.; Stairs, I.H.; Nice, D.J.; Ransom, S.M.; Demorest, P.B.; Arzoumanian, Z.; Crowter, K.; Dolch, T.; et al. The NANOGrav Nine-year Data Set: Mass and Geometric Measurements of Binary Millisecond Pulsars. *Astrophys. J.* **2016**, *832*, 167. [CrossRef]
108. Antoniadis, J.; Freire, P.C.C.; Wex, N.; Tauris, T.M.; Lynch, R.S.; van Kerkwijk, M.H.; Kramer, M.; Bassa, C.; Dhillon, V.S.; Driebe, T.; et al. A Massive Pulsar in a Compact Relativistic Binary. *Science* **2013**, *340*, 6131. . 1233232. [CrossRef] [PubMed]
109. Cromartie, H.T.; Fonseca, E.; Ransom, S.M.; Demorest, P.B.; Arzoumanian, Z.; Blumer, H.; Brook, P.R.; DeCesar, M.E.; Dolch, T.; Ellis, J.A.; et al. Relativistic Shapiro delay measurements of an extremely massive millisecond pulsar. *Nature Astron.* **2019**, *4*, 72–76. [CrossRef]
110. Fonseca, E.; Cromartie, H.T.; Pennucci, T.T.; Ray, P.S.; Kirichenko, A.Y.; Ransom, S.M.; Demorest, P.B.; Stairs, I.H.; Arzoumanian, Z.; Guillemot, L.; et al. Refined Mass and Geometric Measurements of the High-mass PSR J0740+6620. *Astrophys. J. Lett.* **2021**, *915*, L12. [CrossRef]
111. van Kerkwijk, M.H.; Breton, R.; Kulkarni, S.R. Evidence for a Massive Neutron Star from a Radial-Velocity Study of the Companion to the Black Widow Pulsar PSR B1957+20. *Astrophys. J.* **2011**, *728*, 95. [CrossRef]
112. Miller, M.C.; Lamb, F.K.; Dittmann, A.J.; Bogdanov, S.; Arzoumanian, Z.; Gendreau, K.C.; Guillot, S.; Harding, A.K.; Ho, W.C.G.; Lattimer, J.M.; et al. PSR J0030+0451 Mass and Radius from *NICER* Data and Implications for the Properties of Neutron Star Matter. *Astrophys. J. Lett.* **2019**, *887*, L24. [CrossRef]
113. Riley, T.E.; Watts, A.L.; Bogdanov, S.; Ray, P.S.; Ludlam, R.M.; Guillot, S.; Arzoumanian, Z.; Baker, C.L.; Bilous, A.V.; Chakrabarty, D.; et al. A *NICER* View of PSR J0030+0451: Millisecond Pulsar Parameter Estimation. *Astrophys. J. Lett.* **2019**, *887*, L21. [CrossRef]
114. Miller, M.C.; Lamb, F.K.; Dittmann, A.J.; Bogdanov, S.; Arzoumanian, Z.; Gendreau, K.C.; Guillot, S.; Ho, W.C.G.; Lattimer, J.M.; Loewenstein, M.; et al. The Radius of PSR J0740+6620 from NICER and XMM-Newton Data. *Astrophys. J. Lett.* **2021**, *918*, L28. [CrossRef]
115. Riley, T.E.; Watts, A.L.; Ray, P.S.; Bogdanov, S.; Guillot, S.; Morsink, S.M.; Bilous, A.V.; Arzoumanian, Z.; Choudhury, D.; Deneva, J.S.; et al. A NICER View of the Massive Pulsar PSR J0740+6620 Informed by Radio Timing and XMM-Newton Spectroscopy. *Astrophys. J. Lett.* **2021**, *918*, L27. [CrossRef]
116. Abbott, B.P.; Abbott, R.; Abbott, T.D.; Acernese, F.; Ackley, K.; Adams, C.; Adams, T.; Addesso, P.; Adhikari, R.X.; Adya, V.B.; et al. GW170817: Measurements of neutron star radii and equation of state. *Phys. Rev. Lett.* **2018**, *121*, 161101. [CrossRef] [PubMed]
117. Abbott, R.; Abbott, T.D.; Abraham, S.; Acernese, F.; Ackley, K.; Adams, C.; Adhikari, R.X.; Adya, V.B.; Affeldt, C.; Agathos, M.; et al. GW190814: Gravitational Waves from the Coalescence of a 23 Solar Mass Black Hole with a 2.6 Solar Mass Compact Object. *Astrophys. J. Lett.* **2020**, *896*, L44. [CrossRef]
118. Zhang, L.; Huang, M. Holographic cold dense matter constrained by neutron stars. *Phys. Rev. D* **2022**, *106*, 096028. [CrossRef]
119. Karch, A.; Katz, E.; Son, D.T.; Stephanov, M.A. Linear confinement and AdS/QCD. *Phys. Rev. D* **2006**, *74*, 015005. [CrossRef]
120. McLerran, L.; Reddy, S. Quarkyonic Matter and Neutron Stars. *Phys. Rev. Lett.* **2019**, *122*, 122701. [CrossRef]
121. Chen, X.; Li, D.; Hou, D.; Huang, M. Quarkyonic phase from quenched dynamical holographic QCD model. *J. High Energy Phys.* **2020**, *03*, 073. [CrossRef]
122. Deng, W.T.; Huang, X.G. Vorticity in Heavy-Ion Collisions. *Phys. Rev. C* **2016**, *93*, 064907. [CrossRef]
123. Li, H.; Pang, L.G.; Wang, Q.; Xia, X.L. Global Λ polarization in heavy-ion collisions from a transport model. *Phys. Rev. C* **2017**, *96*, 054908. [CrossRef]
124. Wei, D.X.; Deng, W.T.; Huang, X.G. Thermal vorticity and spin polarization in heavy-ion collisions. *Phys. Rev. C* **2019**, *99*, 014905. [CrossRef]
125. Klevansky, S.P.; Lemmer, R.H. Chiral symmetry restoration in the Nambu-Jona-Lasinio model with a constant electromagnetic field. *Phys. Rev. D* **1989**, *39*, 3478–3489. [CrossRef] [PubMed]

126. Klimenko, K.G. Three-dimensional Gross-Neveu model in an external magnetic field. *Teor. Mat. Fiz.* **1991**, *89*, 211–221. [CrossRef]
127. Gusynin, V.P.; Miransky, V.A.; Shovkovy, I.A. Dimensional reduction and catalysis of dynamical symmetry breaking by a magnetic field. *Nucl. Phys. B* **1996**, *462*, 249–290. [CrossRef]
128. Bali, G.S.; Bruckmann, F.; Endrodi, G.; Fodor, Z.; Katz, S.D.; Krieg, S.; Schafer, A.; Szabo, K.K. The QCD phase diagram for external magnetic fields. *J. High Energy Phys.* **2012**, *02*, 044. [CrossRef]
129. Bali, G.S.; Bruckmann, F.; Endrodi, G.; Fodor, Z.; Katz, S.D.; Schafer, A. QCD quark condensate in external magnetic fields. *Phys. Rev. D* **2012**, *86*, 071502. [CrossRef]
130. Bali, G.S.; Bruckmann, F.; Endrodi, G.; Gruber, F.; Schaefer, A. Magnetic field-induced gluonic (inverse) catalysis and pressure (an)isotropy in QCD. *J. High Energy Phys.* **2013**, *04*, 130. [CrossRef]
131. Ding, H.T.; Li, S.T.; Mukherjee, S.; Tomiya, A.; Wang, X.D. Meson masses in external magnetic fields with HISQ fermions. *PoS* **2020**, *LATTICE2019*, 250. [CrossRef]
132. Bali, G.S.; Bruckmann, F.; Constantinou, M.; Costa, M.; Endrodi, G.; Katz, S.D.; Panagopoulos, H.; Schafer, A. Magnetic susceptibility of QCD at zero and at finite temperature from the lattice. *Phys. Rev. D* **2012**, *86*, 094512. [CrossRef]
133. Bali, G.S.; Endrődi, G.; Piemonte, S. Magnetic susceptibility of QCD matter and its decomposition from the lattice. *J. High Energy Phys.* **2020**, *07*, 183. [CrossRef]
134. Xu, K.; Lin, F.; Huang, A.; Huang, M. Λ/Λ^{-} polarization and splitting induced by rotation and magnetic field. *Phys. Rev. D* **2022**, *106*, L071502. [CrossRef]
135. Guo, X.; Liao, J.; Wang, E. Spin Hydrodynamic Generation in the Charged Subatomic Swirl. *Sci. Rep.* **2020**, *10*, 2196. [CrossRef] [PubMed]

symmetry

Article

The Metastable State and the Finite-Size Effect of the First-Order Phase Transition

Mingmei Xu and Yuanfang Wu *

Key Laboratory of Quark and Lepton Physics (MOE) and Institute of Particle Physics, Central China Normal University, Wuhan 430079, China
* Correspondence: wuyf@ccnu.edu.cn

Abstract: In the QCD, a transition restoring the chiral symmetry occurs at a high temperature and density. Searching for the signals of the QCD phase transition is one of the goals of the current relativistic heavy-ion physics programs. The metastable state is a unique feature of the first-order phase transition. Using the van der Waals equation of state, the role of the metastable state in finite-size effects is analyzed. It is found that the finite-size effects of the first-order phase transition are closely related to the metastable state. Metastability can be observed in the distribution of the order parameters and the probability of its occurrence depends on the system scale. A sizable probability of the metastability requires a small enough system size. The possibility of observing the metastability in the RHIC/BES is discussed.

Keywords: relativistic heavy-ion collision; phase transition; finite-size effect

1. Introduction

In the 1970s, T. D. Lee et al. [1,2] predicted that through high-energy heavy-ion collisions, it was possible to form a high-temperature and/or high-density environment so that a new state of matter—quark gluon plasma (QGP) [3,4]—would be produced. Since then, great theoretical and experimental progress has been made in the field. In the 2000s, the relativistic heavy-ion experiments, in particular the Relativistic Heavy-Ion Collider (RHIC) in the Brookheaven National Laboratory with a collision energy of several hundred GeV, had observed the parton (quark and gluon) degrees of freedom [5–8]. It is widely believed that a strongly coupled QGP has been formed. The discovery of the QGP phase is a milestone in heavy-ion physics. However, the phase boundary between the hadron phase and the QGP phase has not been conclusively mapped yet, challenging the theory and experiment.

Consider N_f flavors of massless quarks which couple in the fundamental representation of a $SU(N_c)$ color gauge group. The flavor symmetry of the quantum theory is $G_f = Z_A(N_f) \times SU(N_f) \times SU(N_f)$. At zero temperature, the chromodynamic vacuum spontaneously breaks G_f to $SU(N_f)$, while at a high temperature, the full chiral symmetry is restored to the vacuum [9]. Lattice QCD calculations at a high temperature indicate that the restoration of the spontaneously broken chiral symmetry with a physical quark mass is a smooth crossover at zero baryon chemical potential [10], while several QCD-based models predict a first-order phase transition at a high baryon chemical potential [11–15]. The existence of the critical point, which terminates the first-order phase transition line in the QCD phase diagram, is expected and being searched for in the ongoing heavy-ion experiments.

High cumulants of conserved charges are sensitive observables to the critical point [16–18]. Their sign or the change in the sign can signal the critical point. Especially, the fourth-order cumulant (kurtosis) should show a non-monotonic energy dependence when the chemical freeze-out line gets close to the critical point [19]. The STAR data of the first stage of the beam energy scan (BES) at the RHIC indeed shows the non-monotonic behavior of the kurtosis of the net-proton distribution [20]. However, it is hard to obtain a definite conclusion due to the large error bar, as well as the lack of knowledge on the finite-size effects, the acceptance effect and the non-equilibrium dynamics.

Instead of directly observing the critical point, the identification of the first-order phase transition is thought to be crucial to the determination of the phase boundary. If we see signals of the first-order phase transition, the existence of the critical point is also validated because a first-order phase transition line will definitely terminate at a critical point before touching the crossover region. For that reason, the future second stage of the BES at the RHIC not only enhances the statistics and acceptance at existing energies but also extends the energy range down to lower energies. Other heavy-ion programs, e.g., the Compressed Baryonic Matter Experiment (CBM) at FAIR (in Germany), the Nuclotron-based Ion Collider fAcility (NICA) at JINR (in Russia) and the Cooling Storage Ring (CSR) at HIRFL (in China), will also concentrate on lower energies and hence a larger baryon chemical potential, giving more possibilities to explore the first-order phase transition.

A possible signature of a first-order phase transition is reported in the directed flow measurements in the STAR BES Au+Au collision data [21]. The slope near the mid-rapidity for protons and antiprotons at intermediate impact parameters shows a minimum between a collision energy of 11.5 and 19.6 GeV. The net-proton slope changes the sign twice between 7.7 and 39 GeV, which qualitatively resemble the predictions of a hydrodynamical model with a first-order phase transition. However, the position of the dip in the data differs from that of the hydrodynamic model, and the error bars for different centralities are large, which requires more statistics and a better reaction plane resolution. Therefore, other signals and evidence of the first-order phase transition are still needed.

An interesting feature of the first-order phase transition is the spinodal curve, i.e., the "S"-shape curve of the equation of state, such as the function $p(V)$ of the van der Waals fluid. It refers to an interval of the number density where the derivative of the pressure with respect to the volume is positive, i.e., $(\frac{\partial p}{\partial V})_T > 0$. Such a mechanical instability forms a region on the T-n plane, called the spinodal instable region. It also refers to an interval where $(\frac{\partial p}{\partial V})_T < 0$, but the state is metastable. This region is called the spinodal metastable region. Spinodal instability and metastability are general features of the first-order phase transition and are associated with a convex anomaly in the entropy or free energy [22]. The distribution of the order parameter (number density) which is determined by the free energy also has a convex anomaly, resulting in a double-peak shape.

A double-peak shape of energy and density distributions is demonstrated in a lattice gas model [22]. A similar shape of the density distribution is also shown in a van der Waals fluid [23]. The double-peak distribution can fit the first four cumulants of the proton distribution observed in the STAR data of Au+Au collisions at 7.7 GeV [23]. However, the physical meaning of each peak has not been clarified. Because systems created in heavy-ion collisions are only a few fm in size [24,25], the dependence of the double-peak distribution on the system size, i.e., the role of stable and metastable states in a small-size system, has not been studied so far.

Because the calculation of the QCD phase transition at high baryon chemical potential is unavailable, the van der Waals fluid is used just as an example to demonstrate the general behaviors of the metastable states. This paper is organized as follows. The van der Waals equation of state is given in Section 2. We demonstrate the spinodal metastable region on the T-μ plane in Section 3. The two peaks of the density distribution are analyzed to correspond to the gas and the liquid phase, which appear in the form of a stable state and a metastable state, respectively. There is a notable volume dependence of the probability of the metastable states which is carefully studied. A small system size is favored to observe the metastable state. The metastable states only contribute to finite-size systems. By considering the contribution of the metastable states, some of the finite-size effects of the first-order phase transition are easily understood. A possible metastable state indicating the quark phase in the STAR Au+Au collision data at 7.7 GeV is also shown in Section 3. Section 4 gives the summary.

2. Formalisms and Methods

The standard van der Waals equation of state (EoS) is given by the pressure function in the canonical ensemble (CE) with a fixed particle number as

$$p(V,T,N) = \frac{NT}{V-bN} - a\frac{N^2}{V^2}. \tag{1}$$

p, V, T and N have their usual meanings as the pressure, the volume, the temperature and the particle number, respectively. The parameters a and b describe the attractive interactions and the repulsive interactions caused by the excluded volume.

In order to accommodate systems with varying particle number, we have to extend the van der Waals EoS to the grand canonical ensemble (GCE). Such extending can be found in the pioneering work [26]. For ease of reference, a systematic deduction is given in Appendix A. Relativistic dispersion relation and Boltzmann statistics are used. As a result, the particle number density n ($n \equiv N/V$) as a function of temperature T and chemical potential μ is given by the following transcendental equation,

$$\mu = -T\ln\frac{(1-bn)\varphi(T)}{n} + b\frac{nT}{1-bn} - 2an, \tag{2}$$

with

$$\varphi(T) = \frac{gTm^2}{2\pi^2}K_2\left(\frac{m}{T}\right). \tag{3}$$

Here, g is the degeneracy factor, m is the particle mass and $K_2(m/T)$ denotes the modified Bessel function of the second kind.

In a model of interacting nucleons, $g = 4$ and $m = 938$ MeV are used. We further assume $a = 262$ MeV·fm^3 and $b = 1.11$ fm^3, resulting in a critical point with $T_c = \frac{8a}{27b} = 69.9366$ MeV, $n_c = \frac{1}{3b} = 0.3003$ fm^{-3} and $\mu_c = 791.542$ MeV [23]. The general behavior of the spinodal curve does not change with parameters.

3. Results and Discussions

3.1. Spinodal Metastable Region of the van der Waals Fluid

With the definition $v \equiv V/N$ (volume occupied by each particle), the EoS (1) can transform to

$$p(T,v) = \frac{T}{v-b} - \frac{a}{v^2}. \tag{4}$$

At $T < T_c$, the function $p(v)$ gives an isotherm, i.e., a spinodal curve, shown in Figure 1a.

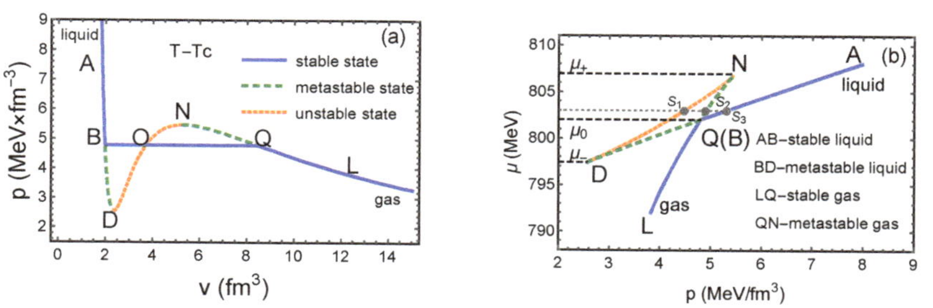

Figure 1. (**a**) An isotherm in $p(v)$ plot at $T = 62$ MeV (a randomly chosen temperature for $T < T_c$). (**b**) An isotherm in $\mu(p)$ plot.

On this curve, the segment DON represents an instable state where the condition of stable equilibrium $(\frac{\partial p}{\partial V})_T < 0$ is violated. An orange dotted line is used to denote unstable states in Figure 1a. The conditions of phase equilibrium require that a horizontal line BOQ is constructed to replace the spinodal curve in order to maintain equal T, p

and μ in the two phases (also known as Maxwell's construction). An equilibrium phase transition between the gas and the liquid takes place along the straight line BOQ rather than the spinodal curve. The systems on segments AB, BOQ and QL correspond to a pure liquid, gas–liquid coexistence and a pure gas, which are all stable states, denoted by blue lines in Figure 1a. The condition of stable equilibrium does not exclude segments BD and NQ. The systems on the two segments are possible to appear but will rapidly evolve to some corresponding states on line BOQ in case of disturbance. That is why they are called metastable states. The metastable states, plotted as green dashed lines in Figure 1a, were confirmed by experiments and named super-heated liquid and super-cooled gas [27], respectively.

Another isotherm can be expressed by $\mu(p)$. By solving the GCE EoS (2) at given T and μ, particle number density n can be obtained. There are unique or three solutions, denoted by $n_i(T,\mu)$, $i = 1$ or $i = 1,2,3$. By putting n_i into the variant form of EoS (1), i.e.,

$$p(T,n) = \frac{nT}{1-bn} - an^2, \tag{5}$$

the pressure $p_i(T,\mu)$ can be obtained, with unique or three values, too. If T is fixed, an isotherm expressed by $\mu(p)$ is plotted in Figure 1b for $T < T_c$. Using the thermodynamical identity $(\frac{\partial\mu}{\partial p})_T = v$, one can present the chemical potential μ_X at any point X on the isotherm as

$$\mu_X = \mu_Y + \int_{p_Y}^{p_X} dp' v(p',T), \tag{6}$$

where the integration is performed along the isotherm from point Y to point X. According to Equation (6), the chemical potential decreases with v if $(\frac{\partial p}{\partial v})_T < 0$, and it increases if $(\frac{\partial p}{\partial v})_T > 0$. Therefore, the chemical potential is a monotonously decreasing function of v along ABD (in Figure 1a) and reaches its minimal value at D. Then, it increases along DON and reaches its maximal value at N. Next, the chemical potential decreases monotonously along NQL. Therefore, segment ABD in Figure 1b is the liquid state, with AB denoting a stable liquid (blue line) and BD denoting a metastable liquid (green dashed line); segment NQL is the gas state, with LQ denoting a stable gas (blue line) and QN denoting a metastable gas (green dashed line); and the coexistence line BOQ in Figure 1a changes to a point Q(B) in Figure 1b.

The chemical potential of the first-order phase transition is denoted as μ_0, i.e., the value at point Q(B) in Figure 1b. μ_- denotes the chemical potential of point D and μ_+ denotes that of point N. As the green dashed lines indicate, metastable states are possible to appear in the interval $\mu \in (\mu_-, \mu_+)$. At another temperature, a similar $\mu(p)$ plot can be obtained, as long as $T < T_c$, giving another set of μ_0, μ_- and μ_+. By such procedure, we obtain the three curves $\mu_0(T)$, $\mu_-(T)$ and $\mu_+(T)$, as plotted in Figure 2.

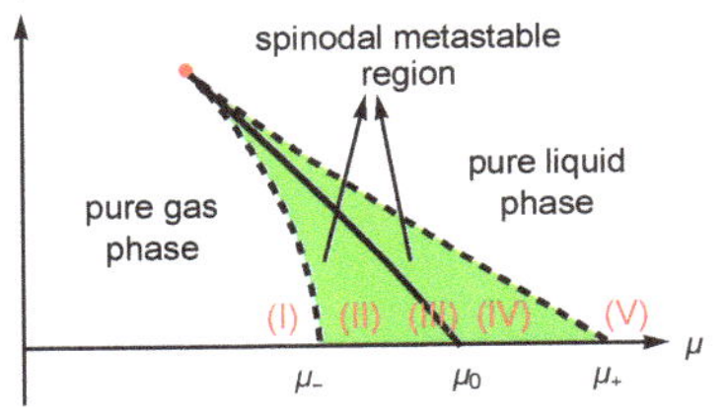

Figure 2. The phase diagram of a van der Waals fluid. The red point denotes the critical point and the solid line denotes the first-order phase transition line in the thermodynamic limit. Spinodal metastable region is marked by the green shaded area.

$\mu_0(T)$, denoted by the solid line in Figure 2, is just the well-known gas–liquid coexistence line, which forms the boundary of gas phase and liquid phase, in the sense of the thermodynamic limit. In Figure 2, the red point denotes the critical point. The region between μ_- and μ_+ is colored in green, demonstrating the spinodal metastable region where both the stable states and the metastable states are possible states. Thus, the phase diagram is divided into five regions, which are marked by Roman numerals I-V in Figure 2. Region I ($\mu < \mu_-$) describes a pure gas; Region II ($\mu_- < \mu < \mu_0$) describes the spinodal region of a stable gas (corresponding to segment LQ in Figure 1b) and a metastable liquid (corresponding to segment BD in Figure 1b); Region III ($\mu = \mu_0$) describes the coexistence of the gas and the liquid; Region IV ($\mu_0 < \mu < \mu_+$) describes the spinodal region of a metastable gas (see segment QN) and a stable liquid (see segment AB); Region V ($\mu > \mu_+$) describes a pure liquid. Both Region II and Region IV are spinodal metastable regions.

In the thermodynamic limit, only the stable states are adopted due to the criteria of stable equilibrium. As Figure 1b shows, in order to choose the stable state, the criteria of the minimum chemical potential at a fixed pressure is equivalent to the criteria of the maximum pressure at a fixed chemical potential. For example, at a given temperature and chemical potential, there are three solutions of Equation (2) in the case of $\mu_- < \mu < \mu_0$ or $\mu_0 < \mu < \mu_+$. The three solutions are located on the orange, green and blue lines, respectively, as points s_1, s_2 and s_3 in Figure 1b show. The solution s_3 on the blue line is chosen due to its maximum pressure. That means the metastable states s_2 do not play a role in the thermodynamic limit. However, in the case of finite volume, metastable states play an important role in the spinodal metastable region, as will be illustrated below.

3.2. The Probability of the Metastable State and Its Dependence on the Volume

Let us show the metastable state in the distribution of the particle number density of the van der Waals fluid and its role in understanding the finite-volume effects. According to the standard formula in statistical physics, the probability for a system with particle number N in the grand canonical ensemble is as follows,

$$P(N) \equiv P(N;V,T,\mu) = \frac{(e^{\frac{\mu}{T}})^N \mathcal{Z}_{\rm CE}^{\rm vdW}(V,T,N)}{\sum_N (e^{\frac{\mu}{T}})^N \mathcal{Z}_{\rm CE}^{\rm vdW}(V,T,N)}, \tag{7}$$

where the partition function of the van der Waals fluid in the canonical ensemble is given in Equation (A7). Based on $P(N)$, the distribution of the number density $P(n)$ can also be obtained, which is volume dependent.

Figure 3 shows the distribution of the number density $P(n)$ in the five regions of the phase diagram.We see double-peak shapes in Regions II, III and IV. In Region III, the two peaks are always of equal height. In Region II, the left peak is dominant, while in Region IV, the right peak is dominant. Four finite volumes are studied in Figure 3. In Regions II and IV, the relative height of the two peaks varies with volume.

The double-peak shape of the distribution is due to a convex anomaly in entropy or free energy [22]. According to the Landau–Ginzburg theory, the free energy has two valleys for $T < T_c$, which results in two peaks in the distribution of the order parameter according to the relation $P(X) \sim \exp(-F)$ [27]. Therefore, the first-order phase transition of finite volume is associated with a double-peak distribution of the order parameter. On the phase boundary, the two peaks are of equal height, while nearby the phase boundary, the two peaks are of different height.

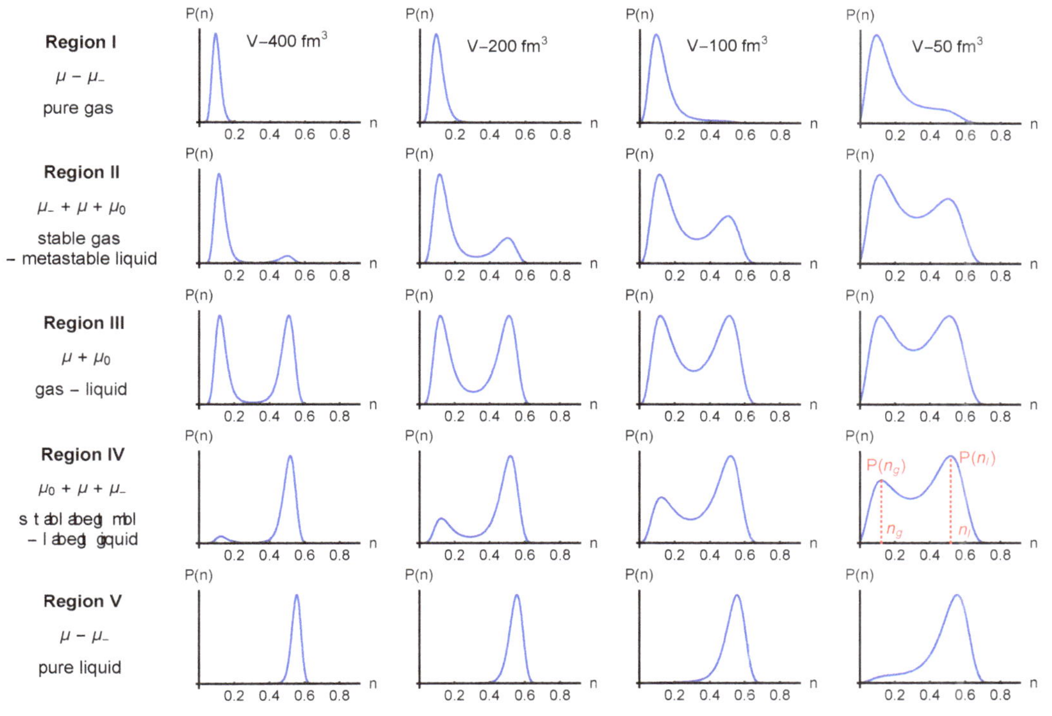

Figure 3. Distribution of the particle number density in five regions of the phase diagram. Each column has the same volume as indicated at the top. From left to right, the volume is 400, 200, 100 and 50 fm^3, respectively. The left peak describes the gas, and the right peak describes the liquid. The higher peak describes the stable state, and the lower peak describes the metastable state.

In order to understand the physical meaning of each peak in Figure 3, we give an example in Table 1. For $T = 62$ MeV and $\mu = 803$ MeV (a randomly chosen point in Region IV), the solutions of Equation (2) are shown in the first column of Table 1. Putting $n_i(i = 1,2,3)$ and $T = 62$ MeV into Equation (5), the corresponding pressure is obtained and shown in the second column of Table 1. A horizontal gray dotted line of $\mu = 803$ MeV is plotted in Figure 1b, showing three intersections with the isotherm. Point s_1 (corresponding to the solution of n_1) is located on the orange dotted line, so n_1 is instable; point s_2 (corresponding to the solution of n_2) is located on the green dashed line QN, so n_2 denotes a metastable gas, i.e., $n_g = n_2 = 0.124643$ fm^{-3}; s_3 (corresponding to the solution n_3) is located on the blue line AB, so n_3 denotes a stable liquid, i.e., $n_l = n_3 = 0.518450$ fm^{-3}. The two data of $n_g = 0.124643$ fm^{-3} and $n_l = 0.518450$ fm^{-3} are drawn as the red dotted lines on one of the distribution plots in Figure 3, which correspond to the maximum of the distribution. The left peak describes the gas, and the right peak describes the liquid. Moreover, the higher peak describes the stable state, and the lower peak describes the metastable state.

Table 1. The solutions of Equation (2) at $T = 62$ MeV and $\mu = 803$ MeV. The three solutions correspond to the three points $s_{1,2,3}$ in Figure 1b.

n (fm^{-3})	p (MeV· fm^{-3})	
$n_1 = 0.290612$	$p_1 = 4.47057$	instable
$n_2 = 0.124643$	$p_2 = 4.89833$	metastable gas ($n_g = n_2$)
$n_3 = 0.518450$	$p_3 = 5.29507$	stable liquid ($n_l = n_3$)

When μ varies from Region I to Region V, the two peaks compete against each other (see from top to bottom in the same column in Figure 3). The unique peaks at Region I and Region V represent a pure gas state and a pure liquid state, respectively. Near the phase boundary, both the gas and the liquid are possible states in the ensemble. The only difference lies in their probabilities of occurrence. On one side of the phase boundary, only one phase is dominant. From Region II to Region IV, the system evolves from a gas-dominant state to a liquid-dominant state.

What do we mean by saying a gas-dominant state nearby the phase boundary? Take the time evolution of the magnetization M in the Monte Carlo simulation of the Ising model below T_c as an example [28]. When on the phase boundary, i.e., the external field $H = 0$, the magnetization M shows many transitions between $+M_L$ and $-M_L$. $M = \pm M_L$ represents two ordered phases, i.e., the upward magnetization and downward magnetization. The time lengths staying at $+M_L$ and $-M_L$ are almost equal. When below the phase boundary, i.e., H obtains a small negative value, M shows transitions between $+M_L$ and $-M_L$ too, with a longer time (larger probability) staying at $-M_L$. If there are many replicas of the system in an ensemble (like the samples obtained by the Monte Carlo simulations), some systems are in the phase of upward magnetization and others are in the phase of downward magnetization. The number of systems of the downward magnetization is more than that of the upward magnetization. That represents a downward magnetization-dominant state. Therefore, a gas-dominant state represents an ensemble where the number of systems staying in a gas state is more than that of liquid.

It is widely known that the phase boundary is well-defined in the thermodynamical limit. However, in the case of finite volume, there can not be a clear boundary. Within the spinodal metastable region, both phases are possible to appear. To quantify the relative probability of the metastable state, we define

$$\zeta \equiv \frac{P_{MS}}{P_{SS}}, \tag{8}$$

where P_{MS} and P_{SS} are the probability peak heights of the metastable state and the stable state, respectively.

The chemical potential dependence of ζ at a fixed volume, e.g., $V = 200$ fm^3, is shown as the blue curve in Figure 4a. ζ equals to 1 at μ_0 and gradually decreases to 0 at both sides. The non-zero relative probability illustrates the contribution of metastable states at a finite volume.

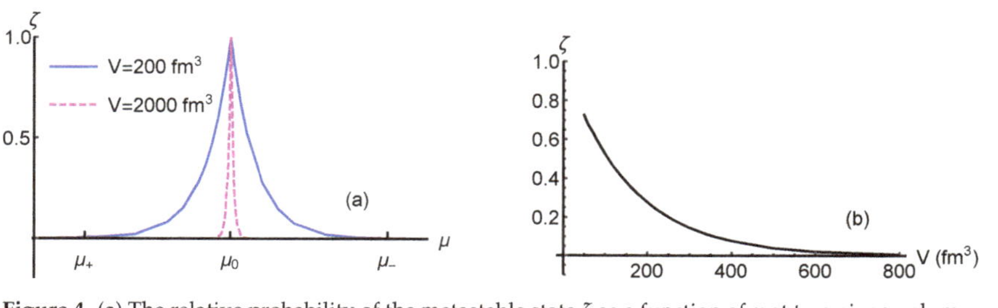

Figure 4. (**a**) The relative probability of the metastable state ζ as a function of μ at two given volumes. (**b**) ζ as a function of volume at $\mu = 803$ MeV ($\mu_0 < \mu < \mu_+$).

As the volume increases to 2000 fm^3, the area enclosed by the purple curve shrinks and ζ decreases to 0 more rapidly. That means a metastable state can be observed within a narrower μ interval. We can expect that the interval of the chemical potential will shrink to zero as the volume approaches infinity. That confirms the statement that metastable states do not have a role to play in the thermodynamic limit.

The spinodal metastable region in the last subsection is specified by the interval (μ_-, μ_+) for systems with a varying particle number. According to Figure 4a, we can infer that the size of the metastable region is volume dependent for finite-size systems. The smaller the volume, the larger the metastable region.

To quantify how the relative height of the two peaks varies with the volume, ζ as a function of volume is shown in Figure 4b. It approximately has a law of

$$\zeta \sim \exp(-V), \tag{9}$$

which is consistent with the Ising model [28].

3.3. Metastable States' Contribution in Understanding Finite-Size Effects of the First-Order Phase Transition

The number density n, as an order parameter of van der Waals fluid, is shown in Figure 5. A spinodal curve ABDNQL is obtained by solving Equation (2) for $T < T_c$. Same as the previous designation, the blue solid line, the green dashed line and the orange dotted line represent stable states, metastable states and instable states, respectively. Because the criteria of a stable equilibrium only chooses stable states, the order parameter n shows the discontinuity in the thermodynamical limit, as the blue line shows in Figure 5.

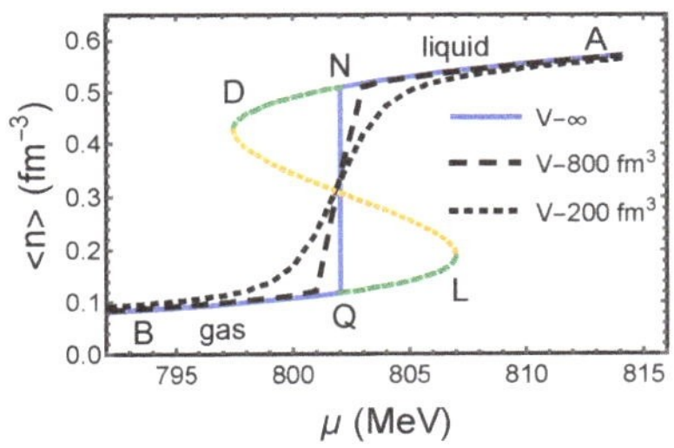

Figure 5. The average number density $\langle n \rangle$ at $T < T_c$.

The number density at finite volume is calculated by obtaining the average from the distribution $P(n)$. The discontinuity of the number density is rounded at finite volumes, as the black dashed line and black dotted line show in Figure 5. The width $\Delta\mu$ over which the transition is rounded is approximately inversely proportional to the volume, i.e.,

$$\Delta\mu \propto \frac{1}{V} = \frac{1}{L^d}, \tag{10}$$

where d is the dimension of the system. This relation is in agreement with the Ising model [28] and the finite-size scaling theory [29–32].

When $\mu < \mu_0$, the ensemble is a gas-dominant state. In this state, the relative probability of the metastable liquid is ζ and the average number density can be approximately related to ζ as a weighted average,

$$\langle n \rangle_V \simeq \frac{n_g + \zeta n_l}{1 + \zeta}. \tag{11}$$

Because ζ is volume-dependent, the average number density is also volume-dependent and can be labeled by a subscript V. Due to $n_g < n_l$, the contribution of the metastable liquid results in $\langle n \rangle_V > n_g$ for $\mu < \mu_0$. That explains that the black curves representing two finite volumes are above the blue curve at the left of the discontinuity point. In particular, due to $\zeta|_{V=200} > \zeta|_{V=800}$ (see the left half of Figure 4a), there is $\langle n \rangle_{V=200} > \langle n \rangle_{V=800}$ for $\mu < \mu_0$. That is the reason why the dotted line (for 200 fm^3) is higher than the dashed line (for 800 fm^3) at the left neighborhood of the discontinuity. Therefore, the volume ordering of the number density at finite volumes reflects the contributions of the metastable states. The smaller the volume, the larger the relative probability of the metastable state, and the smoother and flatter the curve of the number density.

When the chemical potential approaches the discontinuity point μ_0, ζ approaches 1 for whatever volume, as Figure 4a shows. It results in

$$\langle n \rangle_V \bigg|_{\mu_0} \simeq \frac{n_l + n_g}{2}, \quad \text{independent of volume,} \tag{12}$$

which generates the fixed point behavior (the intersection point of curves of different volumes).

3.4. A Possible Metastable State in the STAR Data at 7.7 GeV

The conjectured QCD phase diagram has the same structure as that of the van der Waals fluid. The lattice QCD and NJL models predict that high cumulants of conserved charges are sensitive observables of the critical point [16–18]. The conserved charges, especially the number density of net baryon, denoted as $n_{\rm netB}$, plays the role of the order parameter of the QCD phase transition. It obtains a small value in the hadronic phase and a large value in the quark phase, i.e., $n_{\rm netB}^{\rm hadron} < n_{\rm netB}^{\rm quark}$ [33] (the quark number density shown in Reference [33] is just the baryon number density except a factor of 1/3). Because the freeze-out line is located in the hadronic phase, if it is close to the first-order phase transition line, we hope to see a lower peak on the right side of the distribution representing the metastable quark phase.

In Reference [23], a two-component model was constructed to reproduce the first four factorial cumulants of the proton at 7.7 GeV, especially to explain the large four-particle correlations. If there are two different types of events, denoted by (a) and (b), the distribution of N (the number of proton) is given by

$$P(N) = (1-\alpha)P_{\rm (a)}(N) + \alpha P_{\rm (b)}(N), \tag{13}$$

where $(1-\alpha)$ and α denote the probability that an event belongs to class (a) and (b). $P_{\rm (a)}(N)$ and $P_{\rm (b)}(N)$ are multiplicity distributions governing the event classes (a) and (b), respectively. The factorial cumulants (the relation between the cumulants and the factorial cumulants is discussed in Reference [34]) of the total distribution read

$$\begin{aligned}
\langle N \rangle &= (1-\alpha)\langle N_{\rm (a)} \rangle + \alpha \langle N_{\rm (b)} \rangle, \\
C_2 &= C_2^{\rm (a)} - \alpha(\bar{C}_2 - (1-\alpha)\bar{N}^2), \\
C_3 &= C_3^{\rm (a)} - \alpha\{\bar{C}_3 + (1-\alpha)[(1-2\alpha)\bar{N}^3 - 3\bar{N}\bar{C}_2]\}, \\
C_4 &= C_4^{\rm (a)} - \alpha\{\bar{C}_4 - (1-\alpha)[(1-6\alpha+6\alpha^2)\bar{N}^4 \\
&- 6(1-2\alpha)\bar{N}^2\bar{C}_2 + 4\bar{N}\bar{C}_3 + 3(\bar{C}_2)^2]\},
\end{aligned} \tag{14}$$

where

$$\begin{aligned}
\bar{N} &= \langle N_{\rm (a)} \rangle - \langle N_{\rm (b)} \rangle, \\
\bar{C}_n &= C_n^{(a)} - C_n^{(b)},
\end{aligned} \tag{15}$$

and $C_n^{(a/b)}$ represents the factorial cumulants of the class (a) or (b).

The factorial cumulants of the model are given in Equation (14). The mean value of the first four factorial cumulants in the STAR Au+Au collision data at 7.7 GeV are as follows [20,23],

$$\begin{aligned} \langle N \rangle &= 39.2908,\ C_2 = -1.7511, \\ C_3 &= -10.3098,\ C_4 = 172.906. \end{aligned} \tag{16}$$

A combination of a binomial distribution (event classes (a)) and a Poisson distribution (event classes (b)) are used in Reference [23]. There are four parameters in total, including two parameters from the binomial distribution, one parameter from the Poisson distribution and one weight factor α. One of the parameters in the binomial distribution is fixed and they only fit $\langle N \rangle$, C_3 and C_4. In this section, we follow the method in Reference [23], but fit all the first four factorial cumulants, without fixing any parameters. Each formula in Equation (14) should be equal to the corresponding mean value given in Equation (16), forming four equations. Since the number of parameters is equal to the number of equations, the parameters are exactly determined by solving the set of equations. The result is shown in Figure 6a. The distribution indeed shows a two-peak shape.

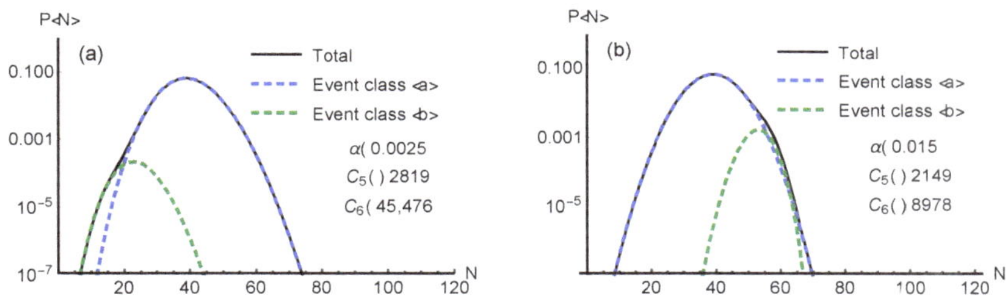

Figure 6. The distribution of the proton number. (**a**) A two-component distribution with a lower peak on the left side, similar to what is performed in Reference [23]. (**b**) A two-component distribution with a lower peak on the right side which is consistent with the scenario of the metastable state. The blue line represents the stable hadron phase, while the green line represents the metastable quark phase.

However, it does not agree with the prediction from the metastable state. Because the number of anti-baryon is much less than that of baryon at low-energy collisions, we expect the number of baryon can approximate the number of net baryon. Thus, $n_{\mathrm{B}}^{\mathrm{hadron}} < n_{\mathrm{B}}^{\mathrm{quark}}$ should hold and $n_{\mathrm{proton}}^{\mathrm{hadron}} < n_{\mathrm{proton}}^{\mathrm{quark}}$ approximately holds. In the case of a first-order phase transition, the lower peak of the metastable state representing the quark phase should be located on the right side. Figure 6a does not show this feature.

In fact, there is not a unique way to identify the distribution in only reproducing the first four factorial cumulants. We try some other fittings and find that a combination of a normal distribution (event class (a)) and a binomial distribution (event class (b)) can also reproduce the first four factorial cumulants measured by STAR, which is shown in Figure 6b. There are five parameters in total, including two parameters from a normal distribution, two from a binomial distribution and one weight factor α. Fitting only four factorial cumulants needs to fix one parameter. Here, we fix the integer parameter of the binomial distribution to be 70. In this case, the lower peak representing the metastable quark phase lies on the right side as the green dashed line in Figure 6b shows, being consistent with the scenario of the first-order phase transition. So far, the physical meaning of the two components becomes clear. They represent two phases. The dominant component (event class (a)) represents the stable phase, and the small component (event class (b)) represents the metastable phase. The presence of the metastable phase may signal a first-order phase transition.

Even though the weight factor α is small ($\alpha = 0.0025$ and 0.015 in Figure 6a,b, respectively), the small component can not be ignored. Without the contribution of the small component, the four factorial cumulants can not be well reproduced. Particularly, the small component has a vital effect on higher cumulants, such as C_6. As the legends show, C_6 = 45,476 and −8978 in Figure 6a,b, respectively. That means, even though the first four factorial cumulants are equal by magnitude for both cases, the fifth factorial cumulant differs by ∼30% and the sixth factorial cumulant differs significantly (by 5 times in magnitude as well as a different sign).

In case the lower peak in Figure 6b is a metastable quark phase, the extremely small weight factor α (reflecting a small relative probability of the metastable state) may hint that the freeze-out point is a little far from the first-order phase transition line. In addition, the two peaks are not separated far enough in the current data. If the collision energy decreases, the temperature decreases and the difference between $n_{\text{proton}}^{\text{hadron}}$ and $n_{\text{proton}}^{\text{quark}}$ increases. The centers of the two peaks will be further apart and it will be better to observe the double-peak structure.

Whether the metastable state is spotted strongly depends on the upcoming measurements of higher factorial cumulants, especially C_6. As indicated in Figure 6, C_5 has the same sign and a similar figure in both cases, while C_6 has an opposite sign and differs much. A definite conclusion should be drawn on the basis of the future C_6 measurement.

4. Summary and Conclusions

In this paper, we revisit the van der Waals fluid and map the spinodal metastable region in the phase diagram. It is a triangular band along the first-order transition line. In the thermodynamic limit, the first-order phase transition line serves as a boundary between the gas and the liquid. However, at a finite volume, the phase boundary loses its original meaning and is replaced by a triangular band where both phases are possible states.

Finite volumes make the metastable state visible in the distribution of the order parameter. On the phase boundary, the distribution has two peaks of equal height. Near the phase boundary, the distribution is still a two-peak shape, but their heights are unequal. The origin of the two-peak distribution is a convex anomaly in the entropy or free energy. The two valleys of free energy at $T < T_c$, as the Landau–Ginzburg theory shows, generate the two peaks of the distribution of the order parameter. The two peaks represent two phases. The higher peak represents the stable state, while the lower peak represents the metastable state. The probabilities of the metastable state are less than that of the stable state. At the left half of the green area in Figure 2, the gas phase is the stable state and its probability in the ensemble is dominant, while the liquid phase is the metastable state and its probability is less prominent. The dominant peak at one side of the phase boundary becomes less dominant at the other side.

The relative probability of the metastable state ζ has significant volume dependence. The rule that the probability ζ decreases exponentially with the volume is consistent with the Ising model. At the same T and μ, the smaller the volume, the larger the probability of the metastable state. As a result, the size of the metastable region is also volume dependent. The smaller the volume, the larger the metastable region in the phase diagram. At a given volume, the closer to the phase boundary, the easier it is to find the metastable state. The metastable states play important roles in understanding the finite-size effects of the first-order phase transition. The smoothness of the discontinuity of the order parameter at finite volume reflects the contribution of the metastable states. The volume ordering, even the fixed point, of the number density can also be understood by considering the contribution of the metastable states.

By fitting the first four factorial cumulants of the proton number in the STAR data at 7.7 GeV, a two-peak distribution, which is consistent with the scenario of the metastable state, is obtained. The presence of a metastable quark phase may hint at a first-order phase transition. However, the statement is still uncertain. Due to the non-uniqueness of the fitting, future measurements of higher factorial cumulants, e.g., C_6, will help to distinguish the scenarios. The fitting made in this study may indicate that the freeze-out point of 7.7 GeV is within the spinodal metastable region. There are two suggestions according to this study. First, the current weight factor α (reflecting the probability of the metastable state) is rather small. To enhance the probability of the metastable state, a smaller system size is favored. Second, the centers of the current two peaks are not separated far enough. A lower collision energy is favored to decrease the temperature. Then, the two peaks will be further apart and easier to observe.

Author Contributions: Conceptualization, M.X. and Y.W.; methodology, M.X.; writing—original draft preparation, M.X.; writing—review and editing, Y.W. All authors have read and agreed to the published version of the manuscript.

Funding: This research was funded by the National Key Research and Development Program of China, grant number 2022YFA1604900, and the National Natural Science Foundation of China, grant number 12275102.

Data Availability Statement: The data used to support the findings of this study are included within the article and are cited at relevant places within the text as references.

Acknowledgments: This paper is dedicated to the Chinese pioneers in the field: Hong-Fang Chen, Lian-Shou Liu, Ru-Keng Su and Qu-Bing Xie. They promoted the development of this field in China. Especially, Lian-Shou Liu was the PhD supervisor of both authors. The authors still remember the days when they had very interesting and memorable discussions with Lian-Shou Liu.

Conflicts of Interest: The authors declare no conflict of interest. The funders had no role in the design of the study; in the collection, analyses or interpretation of data; in the writing of the manuscript; or in the decision to publish the results.

Appendix A. EoS of a van der Waals Fluid in the Grand Canonical Ensemble

Appendix A.1. The Ideal Gas

For an ideal gas, the partition function in the canonical ensemble is

$$\mathcal{Z}_{\rm CE}^{\rm id}(V,T,N) = \int \mathrm{e}^{-\frac{\sum_{i=1}^{N}\sqrt{p_i^2+m^2}}{T}} \frac{1}{N!}\prod_{i=1}^{N}\frac{\mathrm{d}\boldsymbol{x}_i\mathrm{d}\boldsymbol{p}_i}{h^3} = \frac{V^N\varphi^N(T)}{N!}, \tag{A1}$$

where V, T, N have their usual meanings as the volume, the temperature and the particle number. The subscript "CE" denotes the canonical ensemble and the superscript "id" denotes the ideal gas. In Equation (A1),

$$\varphi(T) = \frac{gTm^2}{2\pi^2}K_2\left(\frac{m}{T}\right). \tag{A2}$$

Here, g is the degeneracy factor, m is the particle mass and $K_2(m/T)$ denotes the modified Bessel function of the second kind.

Then, the free energy of the ideal gas reads

$$F_{\rm CE}^{\rm id}(V,T,N) = -T\ln\mathcal{Z}_{\rm CE}^{\rm id} = -NT\left(1+\ln\frac{V\varphi(T)}{N}\right), \tag{A3}$$

with the Stirling's approximation $\ln N! \approx N\ln N - N$ used.

The chemical potential is obtained accordingly,

$$\mu = \left(\frac{\partial F}{\partial N}\right)_{V,T} = -T\ln\frac{V\varphi(T)}{N}. \tag{A4}$$

According to Equation (A4), the number density, defined as $n \equiv N/V$, reads

$$n^{\rm id}(T,\mu) = \varphi(T)e^{\mu/T}. \tag{A5}$$

Appendix A.2. The van der Waals Fluid

The EoS of a van der Waals fluid is usually given by the pressure function as

$$p(V,T,N) = \frac{NT}{V-bN} - a\frac{N^2}{V^2}. \tag{A6}$$

The parameter a describes the attractive interaction and b denotes the repulsive interaction.

The EoS (A6) can be obtained by the mean-field attractive interaction $U = -a\frac{N}{V}$ [27] and the repulsive interaction represented by an excluded volume $V - bN$. Then, its partition function is

$$\begin{aligned} \mathcal{Z}_{\text{CE}}^{\text{vdW}}(V,T,N) &= \int \mathrm{e}^{-\frac{1}{T}(\Sigma_{i=1}^{N}\sqrt{p_i^2+m^2}+UN)} \frac{1}{N!}\prod_{i=1}^{N}\frac{\mathrm{d}\boldsymbol{x}_i\mathrm{d}\boldsymbol{p}_i}{h^3} \\ &= \frac{1}{N!}\varphi^N(T)(V-bN)^N e^{\frac{aN^2}{TV}}. \end{aligned} \tag{A7}$$

Then, its free energy reads

$$F_{\text{CE}}^{\text{vdW}}(V,T,N) = F_{\text{CE}}^{\text{id}}(V-bN,T,N) - a\frac{N^2}{V}, \tag{A8}$$

and the chemical potential will be

$$\mu = -T\ln\frac{(1-bn)\varphi(T)}{n} + b\frac{nT}{1-bn} - 2an. \tag{A9}$$

This is the equation of state of the van der Waals fluid in the grand canonical ensemble. It is a transcendental equation.

References

1. Lee, T.D.; Wick, G.C. Vacuum stability and vacuum excitation in a spin-0 field theory. *Phys. Rev. D* **1974**, *9*, 2291. [CrossRef]

Lee, T.D. Abnormal nuclear states and vacuum excitation. *Rev. Mod. Phys.* **1975**, *47*, 267. [CrossRef]

Collins, J.C.; Perry, M.J. Superdense matter: Neutrons or asymptotically free quarks? *Phys. Rev. Lett.* **1975**, *34*, 1353. [CrossRef]

Shuryak, E.V. Quark-gluon plasma and hadronic production of leptons, photons and psions. *Phys. Lett. B* **1978**, *78*, 150. [CrossRef]

rsene, I.; Bearden, I.G.; Beavis, D.; Besliu, C.; Budick, B.; Bøggild, H.; Chasman, C.; Christensen, C.H.; Christiansen, P.; Cibor, J.; l. Quark-gluon plasma and color glass condensate at RHIC? The perspective from the BRAHMS experiment. *Nucl. Phys. A* *757*, 1–27. [CrossRef]

B.B. (PHOBOS Collab.) The PHOBOS perspective on discoveries at RHIC. *Nucl. Phys. A* **2005**, *757*, 28–101. [CrossRef]

J.; Aggarwal, M.M.; Ahammed, Z.; Amonett, J.; Anderson, B.D.; Arkhipkin, D.; Averichev, G.S.; Badyal, S.K.; Bai, Y.; J.; et al. Experimental and theoretical challenges in the search for the quark-gluon plasma: The STAR Collaboration's ssment of the evidence from RHIC collisions. *Nucl. Phys. A* **2005**, *757*, 102–183. [CrossRef]

ler, S.S.; Afanasiev, S.; Aidala, C.; Ajitanand, N.N.; Akiba, Y.; Al-Jamel, A.; Alexander, J.; Amirikas, R.; Aoki, K.; et al. ense partonic matter in relativistic nucleus-nucleus collisions at RHIC: Experimental evaluation by the PHENIX ucl. Phys. A* **2005**, *757*, 184–283. [CrossRef]

czek, F. Remarks on the chiral phase transition in chromodynamics. *Phys. Rev. D* **1984**, *29* 338. [CrossRef]

.; Fodor, Z.; Katz, S.D.; Szabo, K.K. The order of the quantum chromodynamics transition predicted by the rticle physics. *Nature* **2006**, *443*, 675–678. [CrossRef]

K. Chiral restoration at finite density and temperature. *Nucl. Phys. A* **1989**, *504*, 668–684.

i, R.; Curtis, S.; Gatto, R.; Pettini, G. Chiral phase transitions in QCD for finite temperature and density.). [CrossRef]

or superconductivity and chiral symmetry restoration at non-zero baryon density and temperature. 32. [CrossRef]

hrock, R.E.; Stephanov, M.A.; Verbaarschot, J.J. Phase diagram of QCD. *Phys. Rev. D* **1998**, *58*,

in, I.N.; Rischke, D.H. Chiral phase transition within effective models with constituent quarks. sRef]

uations near the QCD critical point. *Phys. Rev. Lett.* **2009**, *102*, 032301. [CrossRef]

c fluctuations at the QCD phase transition. *Phys. Lett. B* **2006**, *633*, 275–282. [CrossRef]

18. Cheng, M.; Hegde, P.; Jung, C.; Karsch, F.; Kaczmarek, O.; Laermann, E.; Mawhinney, R.D.; Miao, C.; Petreczky, P.; Schmidt, C.; et al. Baryon number, strangeness, and electric charge fluctuations in QCD at high temperature. *Phys. Rev. D* **2009**, *79*, 074505. [CrossRef]
19. Stephanov, M.A. Sign of Kurtosis near the QCD critical point. *Phys. Rev. Lett.* **2011**, *107*, 052301. [CrossRef]
20. Luo, X.; Xu, N. Search for the QCD critical point with fluctuations of conserved quantities in relativistic heavy-ion collisions at RHIC: An overview. *Nucl. Sci. Tech.* **2017**, *28*, 112. [CrossRef]
21. Adamczyk, L.; Adkins, J.K.; Agakishiev, G.; Aggarwal, M.M.; Ahammed, Z.; Alekseev, I.; Alford, J.; Anson, C.D.; Aparin, A.; Arkhipkin, D.; et al. Beam-energy dependence of the directed flow of protons, antiprotons, and pions in Au+Au collisions. *Phys. Rev. Lett.* **2014**, *112*, 162301. [CrossRef] [PubMed]
22. Chomaz, P.; Colonna, M.; Randrup, J. Nuclear spinodal fragmentation. *Phys. Rep.* **2004**, *389*, 263–440. [CrossRef]
23. Bzdak, A.; Koch, V.; Oliinychenko, D.; Steinheimer, J. Large proton cumulants from the superposition of ordinary multiplicity distribution. *Phys. Rev. C* **2018**, *98*, 054901. [CrossRef]
24. Das, S. (for the STAR Collab.) Chemical freeze-out parameters from Au+Au collisions at $\sqrt{s_{NN}} = 7.7$, 11.5 and 39 GeV. *Proc. DAE Symp. Nucl. Phys.* **2011**, *56*, 982.
25. Lacey, R.A. Indications for a critical end point in the phase diagram for hot and dense nuclear matter. *Phys. Rev. Lett.* **2015**, *114*, 142301. [CrossRef]
26. Vovchenko, V.; Anchishkin, D.V.; Gorenstein, M.I. Particle number fluctuations for the van der Waals equation of state. *J. Phys. A Math. Theor.* **2015**, *48*, 305001. [CrossRef]
27. Landau, L.D.; Lifshitz, E.M. *Statistical Physics (Course of Theoretical Physics, Volume 5)*; Pergamon Press: Oxford, UK, 1980.
28. Binder, K.; Landau, D.P. Finite-size scaling at first-order phase transitions. *Phys. Rev. B* **1984**, *30*, 1477. [CrossRef]
29. Fisher, M.E.; Berker, A.N. Scaling for first-order phase transitions in thermodynamic and finite systems. *Phys. Rev. B* **1982**, *26*, 2507. [CrossRef]
30. Binder, K. Theory of first-order phase transitions. *Rep. Prog. Phys.* **1987**, *50*, 783. [CrossRef]
31. van Leeuwen, J.M.J. Singularities in the critical surface and universality for Ising-like spin systems. *Phys. Rev. Lett.* **1975**, *34*, 1056. [CrossRef]
32. Nienhuis, B.; Nauenberg, M. First-order phase transitions in renormalization-group theory. *Phys. Rev. Lett.* **1975**, *35*, 477. [CrossRef]
33. Schaefer, B.J.; Wambach, J. Susceptibilities near the QCD (tri)critical point. *Phys. Rev. D* **2007**, *75*, 085015. [CrossRef]
34. Kitazawa, M.; Luo, X. Properties and uses of factorial cumulants in relativistic heavy-ion collisions. *Phys. Rev. C* **2017**, *96*, 024910. [CrossRef]

symmetry

Article

Collective Excitation in High-Energy Nuclear Collisions—In Memory of Professor Lianshou Liu

Huan Zhong Huang [1], Feng Liu [2], Xiaofeng Luo [2], Shusu Shi [2,*], Fuqiang Wang [3] and Nu Xu [4]

[1] Department of Physics and Astrophysics, University of California, Los Angeles, CA 90095, USA
[2] Key Laboratory of Quark & Lepton Physics (MOE) and Institute of Particle Physics, Central China Normal University, Wuhan 430079, China
[3] Department of Physics, Purdue University, West Lafayette, IN 47907, USA
[4] Nuclear Science Division, Lawrence Berkeley National Laboratory, Berkeley, CA 94720, USA
* Correspondence: shiss@mail.ccnu.edu.cn

Abstract: We celebrate the legacies of our friend and mentor Professor Lianshou Liu who was one of the pioneers for the phenomenology of multi-particle interactions and initiated the physics of relativistic heavy-ion collisions in China. In this article, we discuss some of the recent exciting experimental observations on the collective phenomena including collectivity, chirality, criticality, strangeness production, and thermal equilibrium in high-energy nuclear collisions. Future directions, especially the physics at high baryon density, will be discussed with a focus on the first-order phase boundary and hyperon–nucleon interactions.

Keywords: high-energy nuclear collisions; collectivity; chirality; criticality; QCD; critical point; phase boundary; strangeness; thermalization; viscosity; baryon density

1. Collectivity: Azimuthal Angular Anisotropy in High-Energy Nuclear Collisions

Collective flow defined by the coefficiencies of the Fourier expansion of final particle distribution in momentum space is sensitive to the early stage of nuclear collisions. Specifically, the first three coefficiencies are called directed flow (v_1), elliptic flow (v_2), and triangular flow (v_3), respectively. Directed flow is sensitive to the Equation of State (EoS) of the medium; elliptic flow is sensitive to the degree of freedom, partonic or hadronic level, and degree of equilibrium of the medium; triangular flow is sensitive to the initial geometry fluctuations. A comprehensive set of measurements have been achieved in RHIC-STAR experiment of nuclear collisions [1–9]. The Number of Constituent Quark (NCQ) scaling of v_n observed at high energy collisions (>20 GeV) indicates that the partonic collectivity has been built-up [1–3,8,10]. In particular, the D meson also follows the NCQ scaling [2,10,11], suggesting that the charm quark collectivity is at the same level as that of u, d, and s quarks; therefore, the created medium reaches (nearly) equilibrium.

The main motivation of Beam Energy Scan (BES) program is to explore the QCD phase diagram and search for the possible phase boundary and critical point. The first phase of BES program (BES-I) at STAR experiment covers collision energy $\sqrt{s_{NN}} = 7.7$–62.4 GeV. Lots of interesting phenomena have been observed; here, we focus on the collective flow v_n measurements. Figure 1 summarizes the directed, elliptic, and triangular flow relevant observations from STAR BES-I. The v_1 slope of net-baryon near mid-rapidity as a function of collision energy is regarded as a possible signal of first-order phase transition. The non-monotonic energy dependence of v_1 slope is associated with the phase transition and the minimum of v_1 slope is called the "softest point collapse" [12]. In the experiment, as neutrons are hard to be measured, we use net-proton as a proxy of net-baryon. The left panel of Figure 1 shows the v_1 slope of net-proton, net-Kaon, and net-Λ as a function of collision energy. The non-monotonic behavior is observed for net-proton and net-Λ, and the minimum occurs near $\sqrt{s_{NN}}$ of 10–20 GeV. On the other hand, the net-Kaon slope shows a monotonic increase from low to high energy. Above $\sqrt{s_{NN}} = 20$ GeV, the net-Kaon slope is overlapping with net-proton and net-Λ, while the divergence happens below 20 GeV. Further investigation is needed to understand the physical mechanism of this divergence at the low energy region. In addition, the centrality dependence measurements of net-particle v_1 slope are crucial to understand the effects from mechanism which is not related to phase transition [13–15]. The v_2 of particles and corresponding anti-particles is also found to be significantly different below 20 GeV as shown in the middle panel of Figure 1. Even though the observed NCQ scaling is not valid for particles and anti-particles, roughly scaling still works for particle group and anti-particle group separately. One explanation of the v_2 difference is the transported effect. The final particles are a mixture of transported and produced particles, while the dominant part produces particles at high energy collisions, such as 200 GeV at RHIC and a few TeV at LHC. With the decrease of collision energy, there are more final particles from transportation. Since the transported particles undergo the whole collision evolution, the collectivity should be different with the produced particles. Several model studies try to explain the v_2 difference of particles and anti-particles and succeed partly [16–22]. More detailed measurements as a function of centrality and for (multi-)strange hadrons help us pick out the right mechanism. A similar difference is also observed for v_3, as shown in the right panel of Figure 1. The collision energy and particle type dependence are the same as v_2. Recently, the v_1 and v_2 results of identified particles from $\sqrt{s_{NN}} = 3$ GeV suggest that the hadronic interactions are dominant at such energy [23–25]. Comprehensive measurements of collective flow from collision energy <20 GeV, such as STAR experiment BES-II (3–20 GeV) and experiments at SPS and SIS18 [26–28], help us further constrain the phase transition boundary.

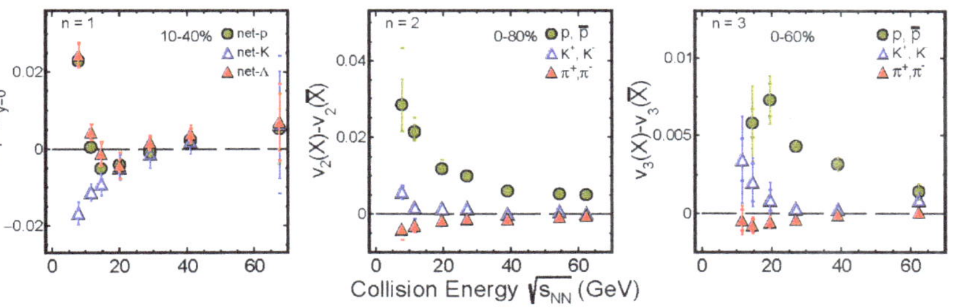

Figure 1. Collision energy dependence of directed flow slope at mid-rapidity (**left**), elliptic flow (**middle**), and triangular flow (**right**) difference of particles versus anti-particles. The data are from [4–9,29].

The energy dependent experimental data on the collective flow can be used to extract important properties of the medium. The temperature dependence of the event-averaged shear viscosity-to-entropy ratio $4\pi\eta/s$ [30] is shown in Figure 2. In the left panel, chemical freeze-out temperature from each energy [31] is used and normalized to that from 200 GeV Au+Au collisions. As one can see, in the high energy limit, $\sqrt{s_{NN}} =$ 39–200 GeV, the ratio reaches unity, the quantum limit, implying that the medium created in such collisions is dominated by partonic interactions with a minimum value of $4\pi\eta/s$. At lower collision energies, on the other hand, hadronic interactions are dominant, and the medium shows a rapid increase of the viscosity-to-entropy ratio. The right panel is taken from Ref. [30], where the temperature evolution of the shear ratio is shown as a function of the scaled temperature T/T_c. Here, T_c represents the critical temperature in the calculation [32,33]. The observed V-shaped feature is quite similar to what is described in Ref. [34] and can be taken as experimental evidence of the expected crossover transition in QCD.

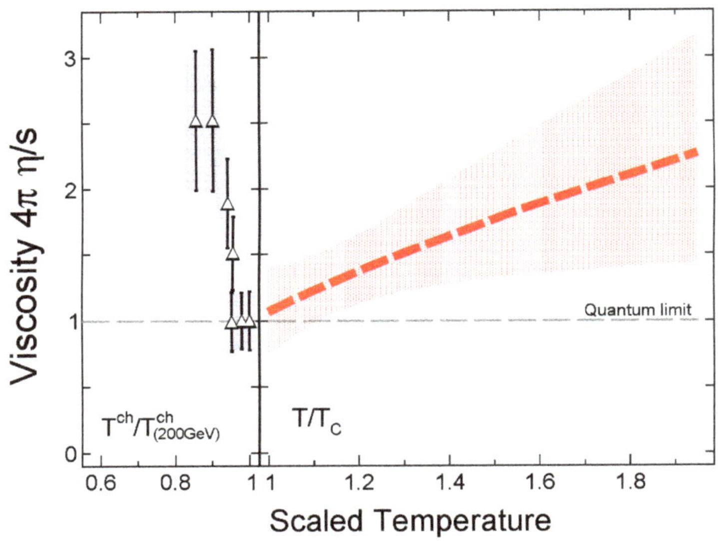

Figure 2. The effective values of shear viscosity-to-entropy ratio, $4\pi\eta/s$, shown as a function of the scaled temperature. The horizontal dashed line indicates the quantum lower limit. Left panel: the extracted $4\pi\eta/s$ from the energy dependence of the measured v_2 [6] and v_3 [35], shown as the scaled chemical freeze-out temperature T_{ch}/T_{ch} (200 GeV). Right panel: temperature evolution of $4\pi\eta/s$, extracted from Bayesian analyses [32,33].

2. Chirality: Chiral Magnetic Effect in High-Energy Nuclear Collisions

The QCD vacuum can exist in states of varying topological charge or Chern–Simons winding number [36–38]. Sphaleron transitions between those states, strongly suppressed under normal conditions, may have appreciable probability at high temperatures, such as those achieved in high-energy nuclear collisions. Local metastable domains of gluon fields with non-zero topological charges may thus form in those collisions [39]. Interactions of quarks with those gluon fields can cause an imbalance in chirality, which may manifest as an electric current along a strong magnetic field, a phenomenon called the Chiral Magnetic Effect (CME) [40]. Because those local domains break the parity ($\mathcal{P}$) and charge–parity ($\mathcal{CP}$) symmetries and because those domains arise ultimately from vacuum fluctuations, out of which our universe is believed to come into being, an observation of the CME in high-energy nuclear collisions may unravel the mystery of the matter–antimatter asymmetry of our universe [41].

The CME has been extensively sought at the RHIC at BNL and the LHC at CERN [42,43]. A signature of the CME is back-to-back charge separation along the magnetic field, produced mainly by spectator protons in relativistic heavy-ion collisions [44]. Because the magnetic field is on average perpendicular to the reaction plane (RP), a commonly used observable is the three-point azimuthal correlator [45],

$$\Delta\gamma = \gamma_{OS} - \gamma_{SS} \quad \text{and} \quad \gamma = \langle \cos(\phi_\alpha + \phi_\beta - 2\psi_{RP}) \rangle \tag{1}$$

where ϕ_α and ϕ_β are the azimuthal angles of particles α and β of either the opposite sin (OS) or the same sign (SS). The ψ_{RP} is the azimuthal angle of the RP and is usually reconstructed from final-state particles, whose inaccuracy is corrected by a resolution factor [46]. Equivalently, γ can also be calculated by the three-particle correlator $\gamma = \langle \cos(\phi_\alpha + \phi_\beta - 2\phi_c) \rangle / v_{2,c}$, where $v_{2,c}$ is the elliptic flow parameter of the third particle c. Several other observables have been proposed and were found to be similar to $\Delta\gamma$ [47].

Charge separation measurements revealed strong $\Delta\gamma$ signals [48–51], approximately independent of collision energy except at lower RHIC energies where the signal dies off [52]. However, major backgrounds exist that arise from two-particle correlations coupled with the finite elliptic flow of those background correlation sources [53–55],

$$\Delta\gamma_{\rm bkgd} = \frac{N_{\rm 2p}}{N_\alpha N_\beta}\langle\cos(\phi_\alpha + \phi_\beta - 2\phi_{\rm 2p})\rangle v_{2,\rm 2p}\,. \tag{2}$$

Here, $N_{\rm 2p}$ and $v_{2,\rm 2p}$ refer, respectively, to the number and elliptic flow of those two-particle correlation sources, such as resonances and jets, and $\phi_{\rm 2p}$ refers to their azimuths; $N_{\alpha,\beta}$ are the α, β particle multiplicities. Experimental indication of major backgrounds comes from small-system measurements [56,57]. Clear experimental evidence of those backgrounds is observed by the STAR measurement of $\Delta\gamma$ as a function of pair invariant mass [58,59].

It became clear because backgrounds are dominant that data-driven methods must be invoked to reliably extract the possible CME signal. An initial attempt was completed by STAR [60] by analyzing data as a function of the event-by-event ellipticity of the particles of interest (i.e., α and β). A similar event-shape engineering (ESE) analysis [61] was carried out by ALICE [62] and CMS [63] by analyzing data as a function of v_2 of the particles of interest at midrapidity in events selected according to the ellipticity in the forward-rapidity region. Data are consistent with vanishing CME signals and upper limits, on the order of 20% of the inclusive $\Delta\gamma$ measurements, have been extracted.

To eliminate large backgrounds, experiments often rely on comparative or relative measurements. One such relative measurement is the isobar $^{44}_{96}$Ru+$^{44}_{96}$Ru and $^{40}_{96}$Zr+$^{40}_{96}$Zr collisions conducted in 2018 at RHIC [64]. Because of the equal mass numbers of the isobar nuclei, the physics backgrounds are expected to be the same; the 10% difference in the atomic numbers is expected to result in a difference in the magnetic fields and thus in the CME signal strengths between the two systems [65]. A blind analysis was conducted [66], and the results [67] are shown in Figure 3 with an unprecedented precision of 0.4%. The Ru+Ru/Zr+Zr ratios of $\Delta\gamma/v_2$ (motivated by Equation (2)) are all below unity, which has been understood to be due mainly to the multiplicity difference between the two isobar systems. This difference was predicted by energy density functional theory calculations [68,69] to arise from the smaller size of the $^{44}_{96}$Ru nucleus compared to that of the $^{40}_{96}$Zr nucleus [70]. If the $\Delta\gamma$ is inversely proportional to multiplicity, then the baseline for the double ratio would be the bottom dashed line in Figure 3. The measured double ratios are all above this line, seemingly suggesting finite CME signals, as also pointed out in Ref. [71]. However, the inverse multiplicity scaling is only approximate because the background in $\Delta\gamma$ scales with the number of correlation sources (see Equation (2)), which may not be strictly proportional to multiplicity. A more realistic baseline may be the Ru+Ru/Zr+Zr ratio of the pair excess $r = (N_{\rm OS} - N_{\rm SS})/N_{\rm OS}$ [67] indicated by the middle dashed line. If so, then there is no evidence of a CME signal in the isobar data. To complicate the matter further, background contamination beyond that of Equation (2) exists because of genuine three-particle correlations and nonflow contamination in v_2 [72]. Preliminary estimates [73] of those nonflow effects are shown by the shaded bands in Figure 3, indicating the final baselines for the double ratios measured by the full event and subevent methods. The isobar data are consistent with these baselines within approximately one standard deviation, suggesting that the CME signal in isobar collisions cannot be larger than a few percent—the 0.4% data precision translates into a $f_{\rm CME}$ = 2–3% CME fraction of the inclusive $\Delta\gamma$ measurement.

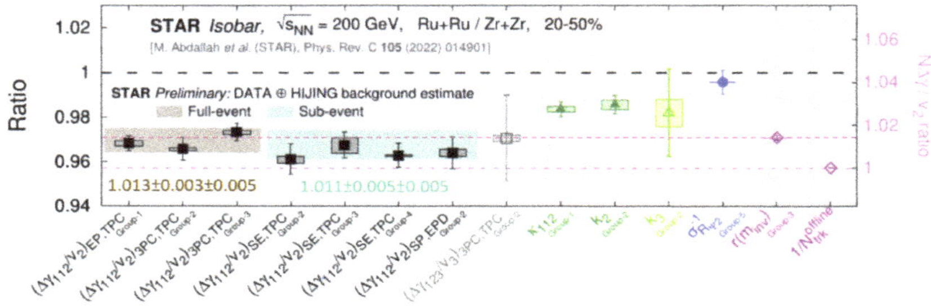

Figure 3. The seven leftmost black data points show the Ru+Ru/Zr+Zr ratios of $\Delta\gamma/v_2$ (left coordinate) and $N\Delta\gamma/v_2$ (right coordinate) from STAR [67]. The bottom and middle dashed lines indicate background baselines if the background scales with inverse multiplicity or relative pair excess $r = (N_{OS} - N_{SS})/N_{OS}$, respectively. The shade bands indicate preliminary estimates of background baselines taking into account effects of nonflow contamination [73–75].

Another comparative method consists of measurements with respect to the spectator plane (SP) and the participant plane (PP) [76,77]. Because the SP is better aligned with the perpendicular orientation of the magnetic field and the PP is determined by the elliptic flow harmonic direction, $\Delta\gamma$ measurements w.r.t. SP and PP contain different contributions from the CME signal and flow background [76]. One can obtain the CME fraction by $f_{\rm CME} \equiv \frac{\Delta\gamma_{\rm CME}\{\rm PP\}}{\Delta\gamma\{\rm PP\}} = \frac{A/a-1}{1/a^2-1}$, where a is the ratio of v_2 measurement w.r.t. SP to that w.r.t. PP, which quantifies the angular spread between SP and PP, and A is the ratio of $\Delta\gamma$ measurement w.r.t. SP to that w.r.t. PP. Figure 4 shows the $f_{\rm CME}$ and the absolute $\Delta\gamma_{\rm CME}$ signal strength in Au+Au collisions at 200 GeV by STAR [78]. While the peripheral 50–80% data are consistent with vanishing CME, the midcentral 20–50% results indicate a finite CME signal of the order of 10%, with a significance ranging from 1 to 3 standard deviations. While the flow-induced background is removed, the results are still contaminated by nonflow contributions. Model estimates [72] of those nonflow contributions indicate a consistent-with-zero contribution to the full-event result and a negative contribution to the subevent results. Estimation utilizing real data is ongoing. If nonflow contamination is small or even negative, then the positive $f_{\rm CME}$ observation by STAR may indeed indicate a finite CME signal. Note this does not contradict the null signal from the isobar data because the signal to background ratio in isobar collisions is expected to be significantly smaller, perhaps by a factor of three [79], than in Au+Au collisions, due to the smaller multiplicity (thus larger background) and the smaller magnetic field strength (thus smaller CME strength) in isobar collisions [79]; the $f_{\rm CME} \sim 10\%$ signal in Au+Au collisions and the less than a few percent signal in isobar collisions are, in fact, consistent with each other.

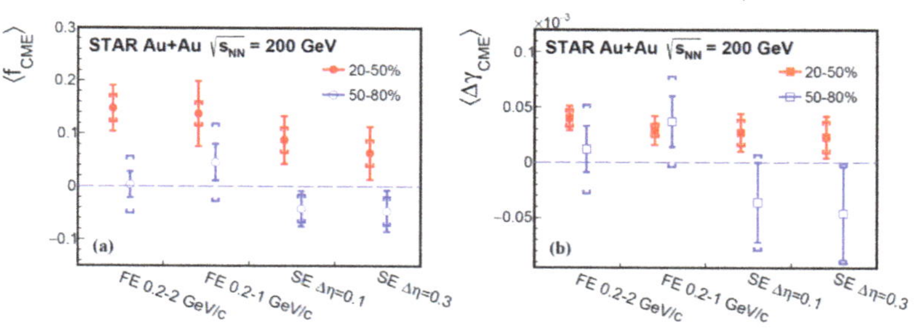

Figure 4. The extracted CME signal fraction $f_{\rm CME}$ (**a**) and magnitude $\Delta\gamma_{\rm CME}$ (**b**) in the inclusive $\Delta\gamma$ measurements by the full-event (FE) and subevent (SE) methods in Au+Au collisions at 200 GeV from STAR [78].

In summary, the CME-sensitive $\Delta\gamma$ measurements in heavy-ion collisions are dominated by flow-induced backgrounds arising from two-particle correlations coupled with elliptic flow v_2. Further background contamination arises from three-particle correlations and nonflow effects in v_2. Methods have been developed to eliminate those backgrounds, including event-shape engineering, isobar collisions, and measurements w.r.t. spectator plane SP and participant plane PP. While the former two have yielded null CME signal with the present statistics, the comparative measurements w.r.t. SP and PP indicate a hint of $f_{\text{CME}} \sim 10\%$ CME signal in Au+Au collisions at 200 GeV, with a significance of 1–3 standard deviations. In the future, an order of magnitude increase in Au+Au statistics is anticipated from 2023 and 2025 by STAR. This would present a powerful data set to either identify the CME or put a stringent upper limit on it. Such an outcome would significantly advance our understanding of QCD and our universe.

3. Criticality: Search for the QCD Critical Point and the Limit of Thermalization in Heavy-Ion Collisions

3.1. High Order Moments and Search for the QCD Critical Point

Exploring the phase structure of hot and dense QCD matter is one of the main goals of heavy-ion collisions. In the large baryon density region, the QCD critical point is the endpoint of the first-order phase transition boundary and a landmark in the QCD phase diagram. Mapping out the first-order phase boundary and pinning down the location of the QCD critical point will enhance our understanding of universe evolution and the structure of visible matters. During the last two decades, many efforts have been made both experimentally [80–91] and theoretically [92–101] to determine the location of the QCD critical point. In this review, we will focus on the measurement of high-order moments with the data taken in the first phase of the beam energy scan program at RHIC.

For a quantum system, the thermodynamic pressure [102] can be expressed via temperature T and chemical potentials μ_i, where $(i = B, Q, S)$ stands for conserved quantum numbers of baryon (B), electric charge (Q), and strangeness (S) in heavy-ion collisions:

$$P(T, \mu_B, \mu_Q, \mu_S, V) = \frac{T}{V}\sum_i \ln Z_i = \sum_i \pm \frac{Tg_i}{2\pi^2}\int k^2 dk \ln\{1 \pm \exp[(\mu_i - E)/T]\}. \tag{3}$$

Here, k, g_i, and $\pm$ signs stand for the particle momentum, the degeneracy, and the nature of Fermion (+) or Boson (−). The sum goes over all particles in the equilibrium with masses up to 2.5 GeV from the PDG.

For a conserved quantity, the cumulants in thermal equilibrium can be expressed through the generalized thermodynamic susceptibilities:

$$C_n^i = \frac{V}{T} T^n \chi_i^{(n)} \tag{4}$$

and the nth order generalized susceptibilities are the derivatives of the pressure:

$$\chi_i^{(n)} = \frac{d^n P}{d\mu_i^n} \tag{5}$$

Fluctuations and correlations among conserved charges, i.e., baryon number (B), electric charge (Q) and strangeness (S), are sensitive observables to probe the QCD phase structure. Experimental proxies, such as net-protons, net-kaons, are used for measurements of mean (M), variance (σ^2), skewness (s), and kurtosis (κ) of conversed charges [82,83,85]. As an example, the connections between moments and cumulant (C_n) ratios of conserved charges are listed as the following:

$$\frac{\sigma_i^2}{M_i} = \frac{C_2^i}{C_1^i} = \frac{\chi_2^i(T,\mu_i)}{\chi_1^i(T,\mu_i)}, \tag{6}$$

$$s_i\,\sigma_i = \frac{C_3^i}{C_2^i} = \frac{\chi_3^i(T,\mu_i)}{\chi_2^i(T,\mu_i)}, \tag{7}$$

$$\kappa_i\,\sigma_i^2 = \frac{C_4^i}{C_2^i} = \frac{\chi_4^i(T,\mu_i)}{\chi_2^i(T,\mu_i)} \tag{8}$$

where $i = B, Q$ and S are the conserved charges, C_n^i is the nth order cumulant, and $\chi_n^i(T,\mu_i)$ is the generalized susceptibility defined as the nth order derivative of pressure with respect to chemical potential μ_i.

One of the most important advantages of Equations (6)–(8) is that they connect the measurements, the experimentally measured moments on the left sides of the equations, to the ratios of the thermodynamic susceptibilities (and ratios of cumulants) from theoretical calculations [98,103–107]. These constructions offer not only a sensitive probe for studying the QCD phase structure in high-energy collisions but also to test the limits of thermalization in such collisions. The latter is a necessary step for us to understand the emerging macroscopic features from the violent microscopic scatterings.

We start first from experimental observations. As a summary, Figure 5 shows recent results on the fourth-order net-proton (filled red circles) and proton (open squares) high moments in central Au+Au collisions from the RHIC BES program and the HADES experiment [86,89,108] together with model comparisons. The thin solid red and dot-dashed blue lines depict a qualitative prediction for the behavior of the fourth-order net-proton cumulant ratio C_4/C_2 ($\kappa\sigma^2$) due to an evolution in the vicinity of a critical region [94]; the locations of the peak of the dot-dashed blue curve and the dip of the solid red curve are chosen only for a qualitative comparison. Intriguing non-monotonic behavior is observed in the data, and while large error bars prevent one from making a more decisive conclusion, the behavior of the fourth-order net-proton cumulant ratio in the energy range $\sqrt{s_{NN}}$ = 7.7–27 GeV seems to significantly deviate from the non-critical baseline provided by models. In particular, both the hadronic transport model UrQMD (cascade mode) [109,110] (gold band) and a thermal model in the canonical ensemble [111] (dot-dashed red line) predict, for decreasing collision energy, a monotonic suppression in the fourth-order cumulant ratio C_4/C_2 due to baryon number conservation, in contrast to the behavior tentatively seen in data. In addition, the experimental data at $\sqrt{s_{NN}} \leq 20$ GeV indicate an excess of two-proton correlations as compared to a non-critical baseline including effects due to, e.g., baryon number conservation [112]. At 3 GeV, the agreement between data and transport model calculation, see the cross in the figure, implies that hadronic interaction dominates the properties of the medium [89,113]. These data imply that the QCD critical region, if created in heavy-ion collisions, could only exist at energies higher than 3 GeV.

The green band, in Figure 5, covers collision energy $\sqrt{s_{NN}}$ = 7.7–19.6 GeV and is the estimated statistical uncertainty from the RHIC beam energy scan phase II program. As one can see, in order to complete the beam energy scan and determine if the QCD critical point exists, the CBM experiment at FAIR is necessary to fill the energy gap between $\sqrt{s_{NN}}$ = 3–8 GeV [114]; see the hatched region in Figure 5. Finally, it is worth noting that, in energy collisions, net-proton cumulants have also been recently proposed as a means extracting the speed of sound and its logarithmic derivative [115].

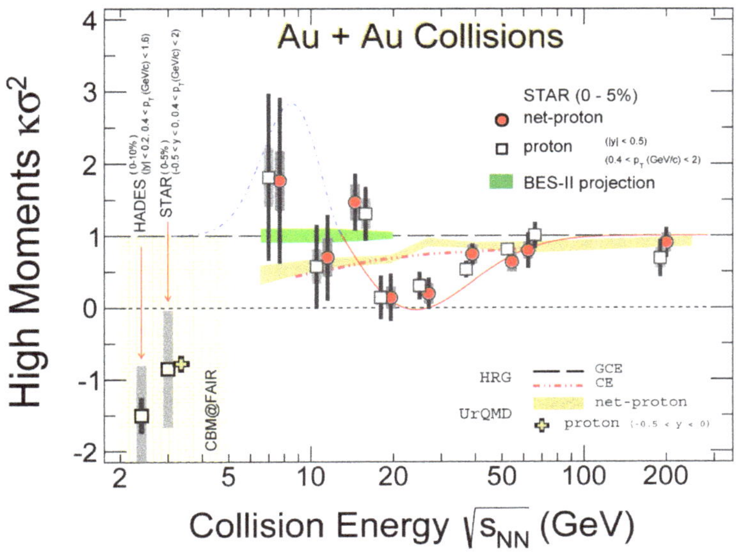

Figure 5. Collision energy dependence of the $\kappa\sigma^2$ (C_4/C_2), for proton (squares) and net-proton (red circles) from top 0–5% Au+Au collisions at RHIC [87,89]. The points for protons are shifted horizontally for clarity. The new result for proton from $\sqrt{s_{NN}}$ = 3 GeV collisions is shown as a filled square. HADES data [108] of $\sqrt{s_{NN}}$ = 2.4 GeV 0–10% collisions are also shown. Statistical uncertainties are presented by bars while the systematic ones are indicated by the vertical grey bands. The green band is the estimated statistical uncertainties from the BES-II program. Results from the Hadron Resonance Gas model [111] and transport model UrQMD [109,110] are shown as the red dot dashed line and gold band, respectively. The thin solid red and dot-dashed blue lines depict a qualitative prediction for the behavior of the fourth-order net-proton cumulant due to an evolution in the vicinity of a critical region [94].

3.2. Limits of Thermalization in High-Energy Nuclear Collisions

In the previous subsection, the energy dependence of the high order moments of net-protons for the search of the QCD critical point in the beam energy scan program at RHIC is discussed. These high order moments, on the other hand, can also be used to test the limits of thermalization in such high-energy collisions [116]. In the past, yields of hadrons from high-energy nuclear collisions have been used to fit to the results of thermal model calculations in order to extract the freeze-out parameters; see, for example, discussions in Ref. [117]. The mean value of hadron yield is the first order of its multiplicity distributions. Certainly, the thermal analysis with the first moment M_i is necessary, but not sufficient, for understanding the dynamics that lead to the macroscopic thermal behavior in high-energy nuclear collisions. Experimentals of high order moments have to be used in the analysis. Below, we report recent progress [118].

In Equations (6)–(8), the terms on the left and in the middle are experimentally measured quantities while the generalized thermodynamic susceptibilities, shown on the right side of these equations, can be extracted from thermodynamic calculations with a given ensemble. By comparing the associated cumulant ratios with experimental data, one then can determine freeze-out temperature and chemical potential. In addition, the test of the limit of thermalization can be performed at any order where data are available. This approach has been used in the tests of Lattice QCD calculations [103–107,119] and in a comparison with experimental results [118].

Here, we present a study using STAR recent published results of net-proton, net-Kaon and net-charge [82,84,85] and compared to a thermal calculation [116] with grand canonical ensemble (GCE). The resulting differences in freeze-out temperature and chemical potential are displayed in Figure 6. Left- and middle-panels show the experimental data of central (0–5%) and peripheral (70–80%) Au+Au collisions. Right-panels display the results of the thermal comparison with central collisions generated by the hadronic transport model UrQMD [109,110]. As one can see, neither peripheral data nor UrQMD central collisions is consistent with the GCE thermal model predictions for Au+Au collisions in the energy range from $\sqrt{s_{NN}}$ = 7.7–200 GeV. For central collisions, on the other hand, the differences in all orders are consistent within 1σ in the high energy range $\sqrt{s_{NN}} \geq 40$ GeV. The results imply that, up to the 3rd order, thermalization is indeed consistent with data in high-energy central heavy ion collisions, but the same cannot be said for collisions below the center of mass energy 40 GeV.

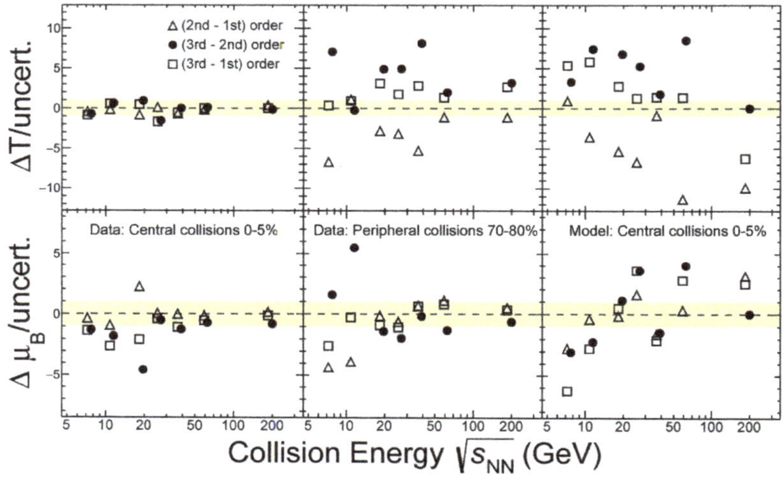

Figure 6. Comparison of the freeze-out parameters obtained by fitting to different orders of high moments of net-proton, net-Kaon, and net-charge from RHIC data [82,84,85]. The differences in freeze-out temperature (ΔT) and baryon chemical potential ($\Delta\mu_B$) are shown in the top and bottom row, respectively. Triangles, dots, and squares represent the difference of net-proton cumulants between 2nd and 1st, 3rd and 2nd, and 3rd and 1st. Left and middle-columns are results from experimental data. The results from hadronic transport model UrQMD [109,110] simulations are shown in the right column. Yellow bands indicate 1σ mark in differences.

The energy dependence of the net-proton fourth-order moment, $\kappa\sigma^2$ from central Au+Au collisions (low panel) are compared with the fitting results (top panel) in Figure 7. Similarly, at high energy, both fitting results shown in the top panel and net-proton kurtosis are consistent with the Grand Canonical limits, and clear deviations appear at energy lower than 40 GeV. In the net-proton kurtosis [86], the expected criticality pattern [94], as a function of energy, seems evident; see the dot-dashed line in the lower panel of Figure 7. It is necessary to point out that criticality is a non-equilibrium phenomenon. The observed deviation from Grand Canonical limit in low energy collisions could be caused by the system passing through the critical point, but it could also be that the system is out of equilibrium. The observation of departures from thermodynamic equilibrium in the final state opens up new directions in the study of heavy-ion collisions. Further analyzing the new experimental precision data [120] with Canonical Ensemble, where the effect of baryon number conservation is included, is necessary in order to understand the underlying physics in the finite baryon density region.

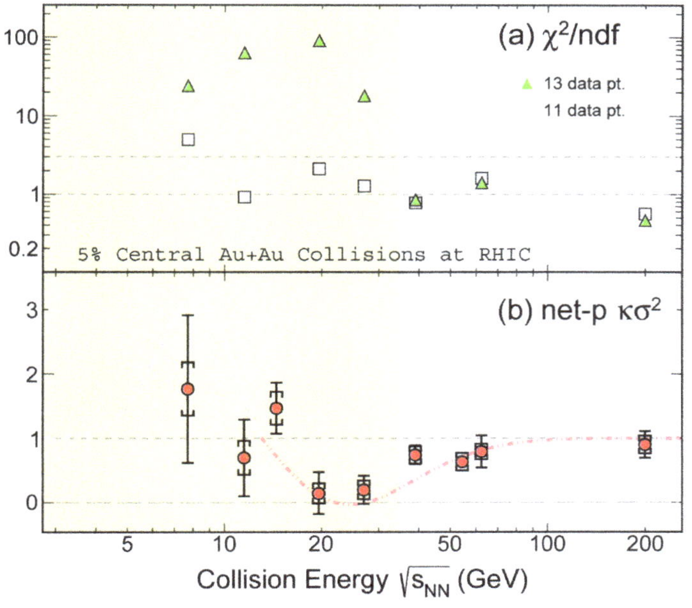

Figure 7. The top panel shows the energy dependence of χ^2/ndf from the thermal fits with either 13 data points (green triangles) or 11 data points (gray squares). The bottom panel shows the energy dependence of the net-proton fourth-order cumulant ratios $C_4/C_2 = \kappa\sigma^2$ [82]. The colour filling in both top and bottom panels tentatively divides the range of collision energies into regions that are clearly in agreement with the predictions of an ideal resonance gas, and therefore cannot contain the QCD critical point. Dot-dashed line is a fit result and indicates possible oscillation in the fourth-order net-proton cumulant ratios [94].

4. Strange Quark Probes of Parton Dynamics and QCD Interactions

Strange quark is a unique probe of the Quark–Gluon Plasma produced in high energy nucleus–nucleus collisions. Rafelski and Muller pointed out [121,122] that, in a hot QGP with a temperature above 160 MeV, the strange quark with a mass of 100 MeV/c^2 can be abundantly produced via gluon–gluon fusion to a strange quark pair. Strangeness enhancement, an increased production of strange hadrons, multi-strange hyperons in particular, has been considered a signature of equilibrated Quark–Gluon Plasma (QGP). On the other hand, if a hadronic gas system can interact and maintain thermal equilibrium hypothetically for a sufficiently long time, strangeness production can also be significantly enhanced to reach an equilibrium. Therefore, it is important to investigate the underlying dynamics for strangeness production.

Strange hadrons can be identified by their decay topologies and such particle identification capability can be extended to momentum much higher than the limit from traditional time-of-flight detectors. Measurements of elliptic flow and nuclear modification factor of identified particles up to the intermediate transverse momentum region of 5 GeV/c have been essential in the discovery of the quark number scaling [123,124]. The quark number scaling is a manifestation of the quark coalescence dynamics for particle formation at the chemical freeze-out stage. The effective degrees of freedom before the hadronization must be dominated by parton dynamics, and the partonic hydrodynamics are responsible for the development of the azimuthal angular anisotropy in the flow measurements. The change of paradigm from a QGP of free quarks and gluons in a QCD bag to a QGP of almost perfect fluid of strongly interacting quarks and gluons is one of the most important achievements of the heavy ion collision physics in the past two decades. The coalescence dynamics showed that the partonic degrees of freedom dominate the evolution dynamics of the QCD matter created in high energy nucleus-nucleus collisions.

The ϕ meson and Ω hyperon have played a unique role in probing the partonic dynamics of the QGP. These strange particles do not have significant hadronic interaction cross sections with other hadrons during the hadronic evolution after the chemical freeze-out. Therefore, they can carry imprints from dynamics of the partonic phase of the QGP evolution. Using coalescence dynamics, these particles have been used to probe strange quark properties in the QGP at the chemical freeze-out [125,126]. Such measurements have also been conducted with the STAR BES data to probe variations of the strange quark properties as the collision energy decreases [127]. We note that the ratio of Ω to $\bar{\Omega}$ increases as the colliding energy decreases from the STAR BES measurement, and the ratio is significantly above one at low energy. There is a net baryon number in Ω hyperons although strangeness conservation dictates that strange and anti-strange quarks must be produced in pairs. The dynamics of baryon number transport to Ω hyperons are a subject of interest. Measurements of correlations between Ω and other particles such as Kaon and anti-hyperons could shed light on the roles of strangeness and baryon number conservation in baryon transport dynamics [128].

The large number of strange hyperons produced in one single heavy ion collision also provided a unique opportunity to investigate hyperon–nucleon, hyperon–hyperon interactions through correlation measurements. The STAR experiment carried out the first meaningful measurement of a Λ–Λ correlation function and extracted the interaction parameters between Λs [129]. Motivated by calculations of the HAL QCD Collaboration [130], STAR measured the p-Ω correlation function from Au+Au collisions [131]. The STAR result, based on the ratio of correlations from two centrality bins as suggested by the theoretical calculation, slightly favors the existence of a bound p-Ω system. ALICE also measured correlations of p-Ξ and p-Ω from $p+p$ collisions at 13 TeV energy. The ALICE data indicated a strong p-Ω attractive interaction though no signal was observed for the existence of a bound state [132]. Existing measurements of hyperon correlations are not precise enough to provide definitive conclusions on the interaction strength between hyperons and the binding energy of possible composite hyperon states. The order of magnitude increase in the collision data samples in the coming years will enable us to utilize the heavy ion collisions as hyperon factories to experimentally address an important topic of hyperon interactions currently incomplete in QCD descriptions of strong interactions.

Because we can identify hyperons at high transverse momentum (p_T) region using their decay topologies, the particle-type (baryon versus meson) dependence of the nuclear modification factors can yield important insight on the parton energy loss in the QGP as well. Figure 8 shows the nuclear modification factors, R_{CP} ratio of normalized yields of central to peripheral collisions, as a function of p_T for charged hadrons, Kaons and Λ hyperons [123]. The R_{CP} values for mesons and hyperons seem to approach each other around p_T of 6 GeV/c. The disappearance of the particle-type dependence could be an important feature of the R_{CP}: it would indicate a minimum p_T above which the jet quenching dynamics may dominate, a landmark p_T cut separating jet probe from hadrodynamical and coalescence regions. Improved measurements of the nuclear modification factors above this important p_T cut of 6 GeV/c for identified particles in the next phase of RHIC scientific program from 2023–2025 could bring new insight on parton energy loss dynamics at RHIC.

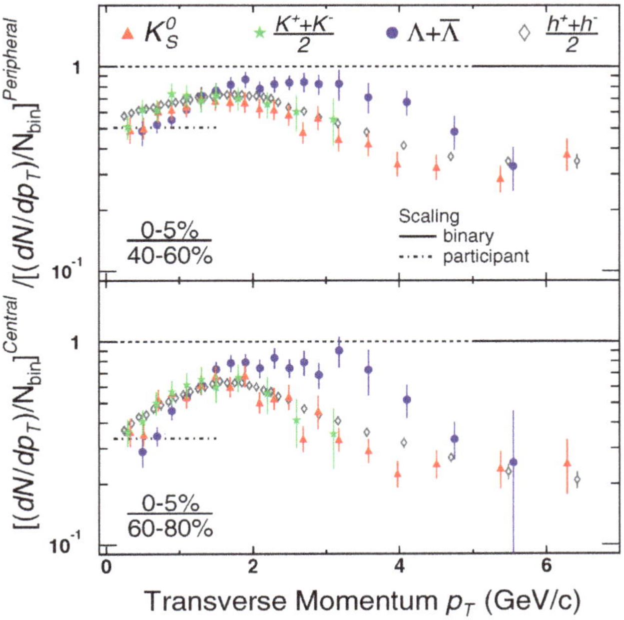

Figure 8. Transverse momentum dependence of the nuclear modification factors for charged hadrons (open diamonds), K^0_S (triangles), charged Kaons (stars) and Λ hyperons (dots) from $\sqrt{s_{NN}} = 200$ GeV Au+Au collisions. Solid and dot-dashed lines stand for the expected scaling of binary collisions and number of participants, respectively. There is a clear particle-type (baryon versus meson) dependence below intermediate p_T presumably due to hydrodynamic flow and coalescence dynamics. The particle-type dependence seems to disappear above p_T of 6 GeV/c where partonic jet quenching dynamics may start to dominate.

Nucleus–nucleus collisions also opened an important venue to study hyper-nuclei dynamics, especially from the STAR BES data to a lower energy regime at future facilities. Driven by baryon density and the strangeness production threshold, both the thermal model [133,134] and transport model [135] have predicted the production of light nuclei and hyper-nuclei to peak around $3 \leq \sqrt{s_{NN}} \leq 10$ GeV in high-energy nuclear collisions. To remove trivial factors including chemical potential and canonical effects, double ratios are often used; for instance, the ratio involving Hypertriton, Helium-3, Λ, and proton yields:

$$S_3 = \frac{^3_\Lambda H/^3He}{\Lambda/p}. \tag{9}$$

The experimental results of S_3 are shown in Figure 9 along with those from model calculations [135–138]. New precision data from STAR experiment show a gradual increase as a function of the collision energy, and the value seems to approach the equilibrium limit in collisions at the LHC [139,140]. It is also interesting to note that the limiting value is about 2/3, which is commonly used in calculations of Λ–N interactions [141]. In the high baryon density region where $\sqrt{s_{NN}} \leq 10$ GeV, the double ratio is further away from the thermal limit, suggesting a clear density effect. More precision measurements are needed in order to understand the Y–N and Y–Y interactions as well as their implication to the inner structure of compact stars. For recent discussions on Y–N and Y–N–N interactions, see Ref. [142] and references therein. The next generation experiment CBM at FAIR with an unprecedentedly interaction rate capability will be very important for such measurements after RHIC.

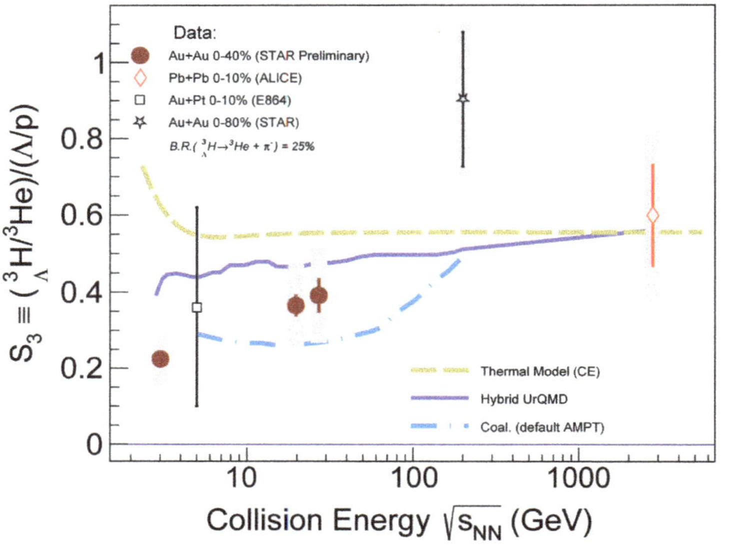

Figure 9. Strangeness population ratio S_3, from central heavy ion collisions, shown as a function of collision energy. Model calculations by thermal model (gold dashed line), hybrid UrQMD (blue line), and AMPT with coalescence (dot-dashed line) are also presented.

5. Outlook: Physics at High Baryon Density

Since the discovery of the new form of matter, the strongly coupled Quark–Gluon Plasma (QGP) [143], created in high-energy nuclear collisions in the early 2000s, scientists have been asking: "What is the structure of the QCD phase diagram in the high baryon density region?" and "Is there a QCD critical point?". Model studies have shown that a first-order phase boundary is expected at the finite baryon chemical potential μ_B, while at vanishing μ_B, there is a smooth crossover between QGP and hadronic matter. Thermodynamically, a critical point ought to be there at the end of the first-order phase boundary; see Figure 10. More discussions on experimental results and Lattice calculations can be found in Refs. [88,99,101,144].

The results, including the observations on collectivity, chirality, critical point, and strangeness production, from the first campaigning of the beam energy scan at RHIC show that partonic activities in central Au+Au collisions persist from $\sqrt{s_{NN}} = 200$ GeV to 39 GeV (corresponding to $\mu_B/T \leq 3$) while in collisions at 3 GeV (corresponding to $\mu_B/T \sim 7$) hadronic interactions dominate [23,86,87,89,124,125]. In addition, charge separation, a measure of the strength of the effect of CME, is vanishing in low energy collisions [52]. The QCD critical point, if existing, should be accessible in collisions at energy between 3–39 GeV. RHIC's second phase energy scan has completed data taking, and more than 10 fold statistics of Au + Au collisions have been collected for energies between 7.7–19.6 GeV. In order to complete the physics program of beam energy scan and study the phase structure of nuclear matter in high baryon density regions, future experiments such as CBM at FAIR are necessary. It will not only aid the search for the QCD critical point, but also extend the research to hyper-nuclear production which is important for studying the basic $Y - N$ interactions. The hyperon–nuclear interaction is one of the vital connections between the physics of high-energy nuclear collisions and inner structure of compact stars.

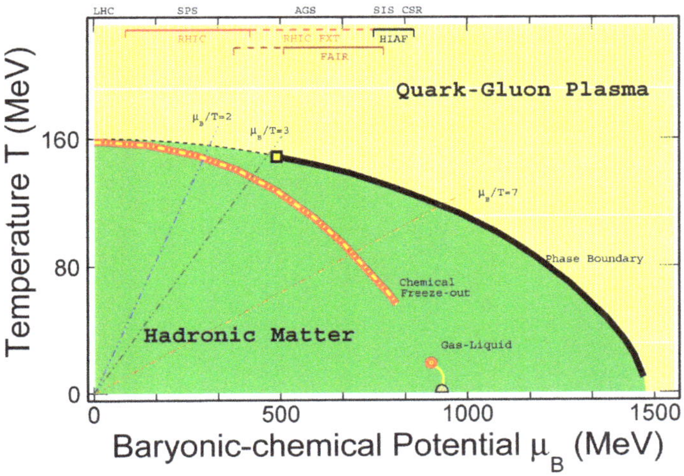

Figure 10. Sketch of the QCD phase diagram. The dashed line represents the smooth crossover region up to $\mu_B/T \leq 3$. The black solid line represents the speculated first-order phase boundary. The empirical thermal freeze-out results from global hadron yield data are shown as the red-yellow line [145]. The liquid-gas transition region that features a second order critical point is shown by the red-circle, and a first-order transition line is shown by the yellow line, which connects the critical point to the ground state of nuclear matter. The coverage of the RHIC BES-II program ($\mu_B/T \sim 3$), STAR fixed target program ($\mu_B/T \sim 7$) and future FAIR and HIAF facilities are indicated at the top of the figure.

Author Contributions: All authors contributed equally. All authors have read and agreed to the published version of the manuscript.

Funding: This work is supported in part by the National Key Research and Development Program of China under contract Nos. 2022YFA1604900, 2020YFE0202002, and 2018YFE0205201; the National Natural Science Foundation of China (NSFC) under contract Nos. 12122505, 12175084, 11890710(11890711), and 11835002; the U.S. Department of Energy (No. DE-SC0012910); and the Fundamental Research Funds for the Central Universities (CCNU220N003).

Data Availability Statement: Not applicable.

Acknowledgments: We thank S. Gupta, V. Koch, D. Mallick, B. Mohanty, and M. Stephanov, for insightful discussions.

Conflicts of Interest: The authors declare no conflict of interest.

References

1. Abdallah, M.S.; Aboona, B.E.; Adam, J.; Adamczyk, L.; Adams, J.R.; Adkins, J.K.; Agakishiev, G.; Aggarwal, I.; Aggarwal, M.M.; Ahammed, Z.; et al. Centrality and transverse momentum dependence of higher-order flow harmonics of identified hadrons in Au+Au collisions at $\sqrt{s_{NN}}$ = 200 GeV. *Phys. Rev. C* **2022**, *105*, 064911. [CrossRef]
2. Adamczyk, L.; Adkins, J.K.; Agakishiev, G.; Aggarwal, M.M.; Ahammed, Z.; Ajitanand, N.N.; Alekseev, I.; Anderson, D.M.; Aoyama, R.; Aparin, A.; et al. Measurement of D^0 Azimuthal Anisotropy at Midrapidity in Au+Au Collisions at $\sqrt{s_{NN}}$ = 200 GeV. *Phys. Rev. Lett.* **2017**, *118*, 212301. [CrossRef]
3. Adamczyk, L.; Adkins, J.K.; Agakishiev, G.; Aggarwal, M.M.; Ahammed, Z.; Alekseev, I.; Aparin, A.; Arkhipkin, D.; Aschenauer, E.C.; Averichev, G.S.; et al. Centrality and transverse momentum dependence of elliptic flow of multistrange hadrons and ϕ meson in Au+Au collisions at $\sqrt{s_{NN}}$ = 200 GeV. *Phys. Rev. Lett.* **2016**, *116*, 062301. [CrossRef]
4. Adamczyk, L.; Adkins, J.K.; Agakishiev, G.; Aggarwal, M.M.; Ahammed, Z.; Alekseev, I.; Alford, J.; Anson, C.D.; Aparin, A.; Arkhipkin, D.; et al. Beam-Energy Dependence of the Directed Flow of Protons, Antiprotons, and Pions in Au+Au Collisions. *Phys. Rev. Lett.* **2014**, *112*, 162301. [CrossRef] [PubMed]

5. Adamczyk, L.; Adams, J.R.; Adkins, J.K.; Agakishiev, G.; Aggarwal, M.M.; Ahammed, Z.; Ajitanand, N.N.; Alekseev, I.; Anderson, D.M.; Aoyama, R.; et al. Beam-Energy Dependence of Directed Flow of Λ, $\bar{\Lambda}$, $K^{\pm}$, K_s^0 and ϕ in Au+Au Collisions. *Phys. Rev. Lett.* **2018**, *120*, 062301. [CrossRef] [PubMed]
6. Adamczyk, L.; Agakishiev, G.; Aggarwal, M.M.; Ahammed, Z.; Alakhverdyants, A.V.; Alekseev, I.; Alford, J.; Anderson, B.D.; Anson, C.D.; Arkhipkin, D.; et al. Inclusive charged hadron elliptic flow in Au + Au collisions at $\sqrt{s_{NN}}$ = 7.7–39 GeV. *Phys. Rev. C* **2012**, *86*, 054908. [CrossRef]
7. Adamczyk, L.; Adkins, J.K.; Agakishiev, G.; Aggarwal, M.M.; Ahammed, Z.; Alekseev, I.; Alford, J.; Anson, C.D.; Aparin, A.; Arkhipkin, D.; et al. Elliptic flow of identified hadrons in Au+Au collisions at $\sqrt{s_{NN}}$ = 7.7–62.4 GeV. *Phys. Rev. C* **2013**, *88*, 014902. [CrossRef]
8. Adamczyk, L.; Adkins, J.K.; Agakishiev, G.; Aggarwal, M.M.; Ahammed, Z.; Alekseev, I.; Alford, J.; Anson, C.D.; Aparin, A.; Arkhipkin, D.; et al. Observation of an Energy-Dependent Difference in Elliptic Flow between Particles and Antiparticles in Relativistic Heavy Ion Collisions. *Phys. Rev. Lett.* **2013**, *110*, 142301. [CrossRef] [PubMed]
9. Adamczyk, L.; Adkins, J.K.; Agakishiev, G.; Aggarwal, M.M.; Ahammed, Z.; Alekseev, I.; Aparin, A.; Arkhipkin, D.; Aschenauer, E.C.; Averichev, G.S.; et al. Centrality dependence of identified particle elliptic flow in relativistic heavy ion collisions at $\sqrt{s_{NN}}$ = 7.7–62.4 GeV. *Phys. Rev. C* **2016**, *93*, 014907. [CrossRef]
10. The ALICE experiment—A journey through QCD. *arXiv* **2022**, arXiv:2211.04384.
11. CMS Collaboration; Sirunyan, A.M.; Bachmair, F.; Bäni, L.; Berger, P.; Bianchini, L.; Casal, B.; Dissertori, G.; Dittmar, M.; Donegà, M.; et al. Measurement of prompt D^0 meson azimuthal anisotropy in Pb-Pb collisions at $\sqrt{s_{NN}}$ = 5.02 TeV. *Phys. Rev. Lett.* **2018**, *120*, 202301. [CrossRef]
12. Stoecker, H. Collective flow signals the quark gluon plasma. *Nucl. Phys. A* **2005**, *750*, 121–147. [CrossRef]
13. Nara, Y.; Jinno, A.; Murase, K.; Ohnishi, A. Directed flow of Λ in high-energy heavy-ion collisions and Λ potential in dense nuclear matter. *Phys. Rev. C* **2022**, *106*, 044902. [CrossRef]
14. Nara, Y.; Ohnishi, A. Mean-field update in the JAM microscopic transport model: Mean-field effects on collective flow in high-energy heavy-ion collisions at $\sqrt{s_{NN}}$ = 2–20 GeV energies. *Phys. Rev. C* **2022**, *105*, 014911. [CrossRef]
15. Nayak, K.; Shi, S.; Xu, N.; Lin, Z.W. Energy dependence study of directed flow in Au+Au collisions using an improved coalescence in a multiphase transport model. *Phys. Rev. C* **2019**, *100*, 054903. [CrossRef]
16. Dunlop, J.C.; Lisa, M.A.; Sorensen, P. Constituent quark scaling violation due to baryon number transport. *Phys. Rev. C* **2011**, *84*, 044914. [CrossRef]
17. Steinheimer, J.; Koch, V.; Bleicher, M. Hydrodynamics at large baryon densities: Understanding proton vs. anti-proton v_2 and other puzzles. *Phys. Rev. C* **2012**, *86*, 044903. [CrossRef]
18. Hatta, Y.; Monnai, A.; Xiao, B.W. Flow harmonics v_n at finite density. *Phys. Rev. D* **2015**, *92*, 114010. [CrossRef]
19. Xu, J.; Song, T.; Ko, C.M.; Li, F. Elliptic flow splitting as a probe of the QCD phase structure at finite baryon chemical potential. *Phys. Rev. Lett.* **2014**, *112*, 012301. [CrossRef] [PubMed]
20. Tu, B.; Shi, S.; Liu, F. Elliptic flow of transported and produced protons in Au+Au collisions with the UrQMD model. *Chin. Phys. C* **2019**, *43*, 054106. [CrossRef]
21. Liu, H.; Wang, F.T.; Sun, K.J.; Xu, J.; Ko, C.M. Isospin splitting of pion elliptic flow in relativistic heavy-ion collisions. *Phys. Lett. B* **2019**, *798*, 135002. [CrossRef]
22. Li, P.; Wang, Y.; Steinheimer, J.; Li, Q.; Zhang, H. Elliptic flow splitting between protons and antiprotons from hadronic potentials. *Mod. Phys. Lett. A* **2020**, *35*, 2050289. [CrossRef]
23. Abdallah, M.; Aboona, B.; Adam, J.; Adamczyk, L.; Adams, J.; Adkins, J.; Agakishiev, G.; Aggarwal, I.; Aggarwal, M.; Ahammed, Z.; et al. Disappearance of partonic collectivity in sNN = 3 GeV Au+Au collisions at RHIC. *Phys. Lett. B* **2022**, *827*, 137003. [CrossRef]
24. Abdallah, M.; Aboona, B.; Adam, J.; Adamczyk, L.; Adams, J.; Adkins, J.; Agakishiev, G.; Aggarwal, I.; Aggarwal, M.; Ahammed, Z.; et al. Light nuclei collectivity from $\sqrt{s_{NN}}$ = 3 GeV Au+Au collisions at RHIC. *Phys. Lett. B* **2022**, *827*, 136941. [CrossRef]
25. Lan, S.W.; Shi, S.S. Anisotropic flow in high baryon density region. *Nucl. Sci. Tech.* **2022**, *33*, 21. [CrossRef]
26. Kuich, M. Highlights from the NA61/SHINE strong-interactions programme. *EPJ Web Conf.* **2022**, *259*, 01001. [CrossRef]
27. Alt, C.; Anticic, T.; Baatar, B.; Barna, D.; Bartke, J.; Behler, M.; Betev, L.; Białkowska, H.; Billmeier, A.; Blume, C.; et al. Directed and elliptic flow of charged pions and protons in Pb + Pb collisions at 40-A-GeV and 158-A-GeV. *Phys. Rev. C* **2003**, *68*, 034903. [CrossRef]
28. Adamczewski-Musch, J.; Arnold, O.; Behnke, C.; Belounnas, A.; Belyaev, A.; Berger-Chen, J.C.; Blanco, A.; Blume, C.; Böhmer, M.; Bordalo, P.; et al. Directed, Elliptic, and Higher Order Flow Harmonics of Protons, Deuterons, and Tritons in Au+Au Collisions at $\sqrt{s_{NN}} = 2.4$ GeV. *Phys. Rev. Lett.* **2020**, *125*, 262301. [CrossRef]
29. Parfenov, P. Elliptic (v_2) and triangular (v_3) anisotropic flow of identified hadrons from the STAR Beam Energy Scan program. *J. Phys. Conf. Ser.* **2020**, *1690*, 012128. [CrossRef]
30. Karpenko, I.A.; Huovinen, P.; Petersen, H.; Bleicher, M. Estimation of the shear viscosity at finite net-baryon density from $A + A$ collision data at $\sqrt{s_{NN}}$ = 7.7–200 GeV. *Phys. Rev. C* **2015**, *91*, 064901. [CrossRef]

31. Adamczyk, L.; Adkins, J.K.; Agakishiev, G.; Aggarwal, M.M.; Ahammed, Z.; Ajitanand, N.N.; Alekseev, I.; Anderson, D.M.; Aoyama, R.; Aparin, A.; et al. Bulk Properties of the Medium Produced in Relativistic Heavy-Ion Collisions from the Beam Energy Scan Program. *Phys. Rev. C* **2017**, *96*, 044904. [CrossRef]
32. Bernhard, J.E.; Moreland, J.S.; Bass, S.A. Bayesian estimation of the specific shear and bulk viscosity of quark–gluon plasma. *Nature Phys.* **2019**, *15*, 1113–1117. [CrossRef]
33. Xu, Y.; Bernhard, J.E.; Bass, S.A.; Nahrgang, M.; Cao, S. Data-driven analysis for the temperature and momentum dependence of the heavy-quark diffusion coefficient in relativistic heavy-ion collisions. *Phys. Rev. C* **2018**, *97*, 014907. [CrossRef]
34. Csernai, L.P.; Kapusta, J.I.; McLerran, L.D. On the Strongly-Interacting Low-Viscosity Matter Created in Relativistic Nuclear Collisions. *Phys. Rev. Lett.* **2006**, *97*, 152303. [CrossRef]
35. Adamczyk, L.; Adkins, J.K.; Agakishiev, G.; Aggarwal, M.M.; Ahammed, Z.; Alekseev, I.; Aparin, A.; Arkhipkin, D.; Aschenauer, E.C.; Attri, A.; et al. Beam Energy Dependence of the Third Harmonic of Azimuthal Correlations in Au+Au Collisions at RHIC. *Phys. Rev. Lett.* **2016**, *116*, 112302. [CrossRef]
36. Lee, T.; Wick, G. Vacuum stability and vacuum excitation in a spin 0 field theory. *Phys.Rev.* **1974**, *D9*, 2291–2316. [CrossRef]
37. Morley, P.D.; Schmidt, I.A. Strong P, CP, T violations in heavy ion collisions. *Z. Phys.* **1985**, *C26*, 627. [CrossRef]
38. Kharzeev, D.; Pisarski, R.; Tytgat, M.H. Possibility of spontaneous parity violation in hot QCD. *Phys. Rev. Lett.* **1998**, *81*, 512–515. [CrossRef]
39. Kharzeev, D. Parity violation in hot QCD: Why it can happen, and how to look for it. *Phys. Lett.* **2006**, *B633*, 260–264. [CrossRef]
40. Fukushima, K.; Kharzeev, D.E.; Warringa, H.J. The chiral magnetic effect. *Phys. Rev.* **2008**, *D78*, 074033. [CrossRef]
41. Dine, M.; Kusenko, A. The Origin of the matter-antimatter asymmetry. *Rev. Mod. Phys.* **2003**, *76*, 1. [CrossRef]
42. Kharzeev, D.E.; Liao, J.; Voloshin, S.A.; Wang, G. Chiral magnetic and vortical effects in high-energy nuclear collisions—A status report. *Prog. Part. Nucl. Phys.* **2016**, *88*, 1–28. [CrossRef]
43. Zhao, J.; Wang, F. Experimental searches for the chiral magnetic effect in heavy-ion collisions. *Prog. Part. Nucl. Phys.* **2019**, *107*, 200–236. [CrossRef]
44. Skokov, V.; Illarionov, A.Y.; Toneev, V. Estimate of the magnetic field strength in heavy-ion collisions. *Int. J. Mod. Phys.* **2009**, *A24*, 5925–5932. [CrossRef]
45. Voloshin, S.A. Parity violation in hot QCD: How to detect it. *Phys. Rev. C* **2004**, *70*, 057901. [CrossRef]
46. Poskanzer, A.M.; Voloshin, S. Methods for analyzing anisotropic flow in relativistic nuclear collisions. *Phys. Rev.* **1998**, *C58*, 1671–1678. [CrossRef]
47. Choudhury, S.; Dong, X.; Drachenberg, J.; Dunlop, J.; Esumi, S.; Feng, Y.; Finch, E.; Hu, Y.; Jia, J.; Lauret, J.; et al. Investigation of experimental observables in search of the chiral magnetic effect in heavy-ion collisions in the STAR experiment. *Chin. Phys. C* **2022**, *46*, 014101. [CrossRef]
48. Abelev, B.I.; Aggarwal, M.M.; Ahammed, Z.; Alakhverdyants, A.V.; Anderson, B.D.; Arkhipkin, D.; Averichev, G.S.; Balewski, J.; Barannikova, O.; Barnby, L.S.; et al. Azimuthal Charged-Particle Correlations and Possible Local Strong Parity Violation. *Phys. Rev. Lett.* **2009**, *103*, 251601. [CrossRef]
49. Abelev, B.I.; Aggarwal, M.M.; Ahammed, Z.; Alakhverdyants, A.V.; Anderson, B.D.; Arkhipkin, D.; Averichev, G.S.; Balewski, J.; Barannikova, O.; Barnby, L.S.; et al. Observation of charge-dependent azimuthal correlations and possible local strong parity violation in heavy ion collisions. *Phys. Rev.* **2010**, *C81*, 054908. [CrossRef]
50. Abelev, B.; Adam, J.; Adamová, D.; Adare, A.M.; Aggarwal, M.M.; Rinella, G.A.; Agocs, A.G.; Agostinelli, A.; Salazar, S.A.; Ahammed, Z.; et al. Charge separation relative to the reaction plane in Pb-Pb collisions at $\sqrt{s_{NN}}$ = 2.76 TeV. *Phys. Rev. Lett.* **2013**, *110*, 012301. [CrossRef]
51. Adamczyk, L.; Adkins, J.K.; Agakishiev, G.; Aggarwal, M.M.; Ahammed, Z.; Alekseev, I.; Alford, J.; Anson, C.D.; Aparin, A.; Arkhipkin, D.; et al. Fluctuations of charge separation perpendicular to the event plane and local parity violation in $\sqrt{s_{NN}}$ = 200 GeV Au+Au collisions at the BNL Relativistic Heavy Ion Collider. *Phys. Rev.* **2013**, *C88*, 064911. [CrossRef]
52. Adamczyk, L.; Adkins, J.K.; Agakishiev, G.; Aggarwal, M.M.; Ahammed, Z.; Alekseev, I.; Alford, J.; Anson, C.D.; Aparin, A.; Arkhipkin, D.; et al. Beam-energy dependence of charge separation along the magnetic field in Au+Au collisions at RHIC. *Phys. Rev. Lett.* **2014**, *113*, 052302. [CrossRef] [PubMed]
53. Wang, F. Effects of Cluster Particle Correlations on Local Parity Violation Observables. *Phys. Rev.* **2010**, *C81*, 064902. [CrossRef]
54. Bzdak, A.; Koch, V.; Liao, J. Remarks on possible local parity violation in heavy ion collisions. *Phys. Rev.* **2010**, *C81*, 031901. [CrossRef]
55. Schlichting, S.; Pratt, S. Charge conservation at energies available at the BNL Relativistic Heavy Ion Collider and contributions to local parity violation observables. *Phys. Rev.* **2011**, *C83*, 014913. [CrossRef]
56. Khachatryan, V.; Sirunyan, A.M.; Tumasyan, A.; Adam, W.; Asilar, E.; Bergauer, T.; Brandstetter, J.; Brondolin, E.; Dragicevic, M.; Erö, J.; et al. Observation of charge-dependent azimuthal correlations in p-Pb collisions and its implication for the search for the chiral magnetic effect. *Phys. Rev. Lett.* **2017**, *118*, 122301. [CrossRef]
57. Adam, J.; Adamczyk, L.; Adams, J.; Adkins, J.; Agakishiev, G.; Aggarwal, M.; Ahammed, Z.; Alekseev, I.; Anderson, D.; Aoyama, R.; et al. Charge-dependent pair correlations relative to a third particle in p + Au and d + Au collisions at RHIC. *Phys. Lett.* **2019**, *B798*, 134975. [CrossRef]
58. Zhao, J.; Li, H.; Wang, F. Isolating the chiral magnetic effect from backgrounds by pair invariant mass. *Eur. Phys. J.* **2019**, *C79*, 168. [CrossRef]

59. Abdallah, M.S.; Adam, J.; Adamczyk, L.; Adams, J.R.; Adkins, J.K.; Agakishiev, G.; Aggarwal, I.; Aggarwal, M.M.; Ahammed, Z.; Alekseev, I.; et al. Pair invariant mass to isolate background in the search for the chiral magnetic effect in Au + Au collisions at sNN = 200 GeV. *Phys. Rev. C* **2022**, *106*, 034908. [CrossRef]
60. Adamczyk, L.; Adkins, J.K.; Agakishiev, G.; Aggarwal, M.M.; Ahammed, Z.; Alakhverdyants, A.V.; Alekseev, I.; Alford, J.; Anson, C.D.; Arkhipkin, D.; et al. Measurement of charge multiplicity asymmetry correlations in high-energy nucleus-nucleus collisions at $\sqrt{s_{NN}} = 200$ GeV. *Phys. Rev.* **2014**, *C89*, 044908. [CrossRef]
61. Schukraft, J.; Timmins, A.; Voloshin, S.A. Ultra-relativistic nuclear collisions: Event shape engineering. *Phys. Lett.* **2013**, *B719*, 394–398. [CrossRef]
62. Acharya, S.; Adam, J.; Adamová, D.; Adolfsson, J.; Aggarwal, M.; Rinella, G.A.; Agnello, M.; Agrawal, N.; Ahammed, Z.; Ahmad, N.; et al. Constraining the magnitude of the Chiral Magnetic Effect with Event Shape Engineering in Pb-Pb collisions at $\sqrt{s_{NN}} = 2.76$ TeV. *Phys. Lett.* **2018**, *B777*, 151–162. [CrossRef]
63. Sirunyan, A.M.; Tumasyan, A.; Adam, W.; Ambrogi, F.; Asilar, E.; Bergauer, T.; Brandstetter, J.; Brondolin, E.; Dragicevic, M.; Erö, J.; et al. Constraints on the chiral magnetic effect using charge-dependent azimuthal correlations in *p*Pb and PbPb collisions at the CERN Large Hadron Collider. *Phys. Rev.* **2018**, *C97*, 044912. [CrossRef]
64. Koch, V.; Schlichting, S.; Skokov, V.; Sorensen, P.; Thomas, J.; Voloshin, S.; Wang, G.; Yee, H.-U. Status of the chiral magnetic effect and collisions of isobars. *Chin. Phys.* **2017**, *C41*, 072001. [CrossRef]
65. Voloshin, S.A. Testing the chiral magnetic effect with central U+U collisions. *Phys. Rev. Lett.* **2010**, *105*, 172301. [CrossRef] [PubMed]
66. Adam, J.; Adamczyk, L.; Adams, J.R.; Adkins, J.K.; Agakishiev, G.; Aggarwal, M.M.; Ahammed, Z.; Alekseev, I.; Anderson, D.M.; Aparin, A.; et al. Methods for a blind analysis of isobar data collected by the STAR collaboration. *Nucl. Sci. Tech.* **2021**, *32*, 48. [CrossRef]
67. Abdallah, M.S.; Aboona, B.E.; Adam, J.; Adamczyk, L.; Adams, J.R.; Adkins, J.K.; Agakishiev, G.; Aggarwal, I.; Aggarwal, M.M.; Ahammed, Z.; et al. Search for the chiral magnetic effect with isobar collisions at $\sqrt{s_{NN}} = 200$ GeV by the STAR Collaboration at the BNL Relativistic Heavy Ion Collider. *Phys. Rev. C* **2022**, *105*, 014901. [CrossRef]
68. Xu, H.-J.; Wang, X.; Li, H.; Zhao, J.; Lin, Z.-W.; Shen, C.; Wang, F. Importance of isobar density distributions on the chiral magnetic effect search. *Phys. Rev. Lett.* **2018**, *121*, 022301. [CrossRef] [PubMed]
69. Li, H.; Xu, H.-J.; Zhao, J.; Lin, Z.-W.; Zhang, H.; Wang, X.; Shen, C.; Wang, F. Multiphase transport model predictions of isobaric collisions with nuclear structure from density functional theory. *Phys. Rev.* **2018**, *C98*, 054907. [CrossRef]
70. Xu, H.-J.; Li, H.; Wang, X.; Shen, C.; Wang, F. Determine the neutron skin type by relativistic isobaric collisions. *Phys. Lett. B* **2021**, *819*, 136453. [CrossRef]
71. Kharzeev, D.E.; Liao, J.; Shi, S. Implications of the isobar-run results for the chiral magnetic effect in heavy-ion collisions. *Phys. Rev. C* **2022**, *106*, L051903. [CrossRef]
72. Feng, Y.; Zhao, J.; Li, H.; Xu, H.j.; Wang, F. Two- and three-particle nonflow contributions to the chiral magnetic effect measurement by spectator and participant planes in relativistic heavy ion collisions. *Phys. Rev. C* **2022**, *105*, 024913. [CrossRef]
73. Feng, Y. Estimate of a nonflow baseline for the chiral magnetic effect in isobar collisions at RHIC. In Proceedings of the 20th International Conference on Strangeness in Quark Matter 2022, Busan, Republic of Korea, 13–17 June 2022.
74. Tribedy, P. STAR Hightlights. In Proceedings of the 29th International Conference on Ultra-Relativistic Nucleus-Nucleus Collisions, Krakow, Poland, 4–10 April 2022.
75. Wang, F. CME–Experimental Results and Interpretation. In Proceedings of the 29th International Conference on Ultra-Relativistic Nucleus-Nucleus Collisions, Krakow, Poland, 4–10 April 2022.
76. Xu, H.-J.; Zhao, J.; Wang, X.-B.; Li, H.-L.; Lin, Z.-W.; Shen, C.-W.; Wang, F.-Q. Varying the chiral magnetic effect relative to flow in a single nucleus-nucleus collision. *Chin. Phys.* **2018**, *C42*, 084103. [CrossRef]
77. Voloshin, S.A. Estimate of the signal from the chiral magnetic effect in heavy-ion collisions from measurements relative to the participant and spectator flow planes. *Phys. Rev.* **2018**, *C98*, 054911. [CrossRef]
78. Abdallah, M.S.; Adam, J.; Adamczyk, L.; Adams, J.R.; Adkins, J.K.; Agakishiev, G.; Aggarwal, I.; Aggarwal, M.M.; Ahammed, Z.; Alekseev, I.; et al. Search for the Chiral Magnetic Effect via Charge-Dependent Azimuthal Correlations Relative to Spectator and Participant Planes in Au+Au Collisions at $\sqrt{s_{NN}} = 200$ GeV. *Phys. Rev. Lett.* **2022**, *128*, 092301. [CrossRef]
79. Feng, Y.; Lin, Y.; Zhao, J.; Wang, F. Revisit the chiral magnetic effect expectation in isobaric collisions at the relativistic heavy ion collider. *Phys. Lett. B* **2021**, *820*, 136549. [CrossRef]
80. Anticic, T.; Baatar, B.; Bartke, J.; Beck, H.; Betev, L.; Białkowska, H.; Blume, C.; Bogusz, M.; Boimska, B.; Book, J.; et al. Critical fluctuations of the proton density in A+A collisions at 158*A* GeV. *Eur. Phys. J. C* **2015**, *75*, 587. [CrossRef]
81. Czopowicz, T. Search for critical point via intermittency analysis in NA61/SHINE. *PoS* **2022**, *400*, 039. [CrossRef]
82. Aggarwal, M.M.; Ahammed, Z.; Alakhverdyants, A.V.; Alekseev, I.; Alford, J.; Anderson, B.D.; Arkhipkin, D.; Averichev, G.S.; Balewski, J.; Barnby, L.S.; et al. Higher Moments of Net-proton Multiplicity Distributions at RHIC. *Phys. Rev. Lett.* **2010**, *105*, 022302. [CrossRef] [PubMed]
83. Adamczyk, L.; Adkins, J.K.; Agakishiev, G.; Aggarwal, M.M.; Ahammed, Z.; Alekseev, I.; Alford, J.; Anson, C.D.; Aparin, A.; Arkhipkin, D.; et al. Energy Dependence of Moments of Net-proton Multiplicity Distributions at RHIC. *Phys. Rev. Lett.* **2014**, *112*, 032302. [CrossRef] [PubMed]

84. Adamczyk, L.; Adkins, J.K.; Agakishiev, G.; Aggarwal, M.M.; Ahammed, Z.; Alekseev, I.; Alford, J.; Anson, C.D.; Aparin, A.; Arkhipkin, D.; et al. Beam energy dependence of moments of the net-charge multiplicity distributions in Au+Au collisions at RHIC. *Phys. Rev. Lett.* **2014**, *113*, 092301. [CrossRef]
85. Adamczyk, L.; Adams, J.; Adkins, J.; Agakishiev, G.; Aggarwal, M.; Ahammed, Z.; Ajitanand, N.; Alekseev, I.; Anderson, D.; Aoyama, R.; et al. Collision Energy Dependence of Moments of Net-Kaon Multiplicity Distributions at RHIC. *Phys. Lett. B* **2018**, *785*, 551–560. [CrossRef]
86. Adam, J.; Adamczyk, L.; Adams, J.R.; Adkins, J.K.; Agakishiev, G.; Aggarwal, M.M.; Ahammed, Z.; Alekseev, I.; Anderson, D.M.; Aparin, A.; et al. Nonmonotonic Energy Dependence of Net-Proton Number Fluctuations. *Phys. Rev. Lett.* **2021**, *126*, 092301. [CrossRef] [PubMed]
87. Abdallah, M.S.; Adam, J.; Adamczyk, L.; Adams, J.R.; Adkins, J.K.; Agakishiev, G.; Aggarwal, I.; Aggarwal, M.M.; Ahammed, Z.; Alekseev, I.; et al. Cumulants and correlation functions of net-proton, proton, and antiproton multiplicity distributions in Au+Au collisions at energies available at the BNL Relativistic Heavy Ion Collider. *Phys. Rev. C* **2021**, *104*, 024902. [CrossRef]
88. Abdallah, M.S.; Adam, J.; Adamczyk, L.; Adams, J.R.; Adkins, J.K.; Agakishiev, G.; Aggarwal, I.; Aggarwal, M.M.; Ahammed, Z.; Alekseev, I.; et al. Measurement of the sixth-order cumulant of net-proton multiplicity distributions in Au+Au collisions at $\sqrt{s_{NN}}$ = 27, 54.4, and 200 GeV at RHIC. *Phys. Rev. Lett.* **2021**, *127*, 262301. [CrossRef]
89. Abdallah, M.S.; Aboona, B.E.; Adam, J.; Adamczyk, L.; Adams, J.R.; Adkins, J.K.; Agakishiev, G.; Aggarwal, I.; Aggarwal, M.M.; Ahammed, Z.; et al. Measurements of Proton High Order Cumulants in $\sqrt{s_{NN}} = 3$ GeV Au+Au Collisions and Implications for the QCD Critical Point. *Phys. Rev. Lett.* **2022**, *128*, 202303. [CrossRef]
90. Beam Energy Dependence of Triton Production and Yield Ratio ($N_t \times N_p/N_d^2$) in Au+Au Collisions at RHIC. *arXiv* **2022**, arXiv:2209.08058.
91. Beam Energy Dependence of Fifth and Sixth-Order Net-proton Number Fluctuations in Au+Au Collisions at RHIC. *arXiv* **2022**, arXiv:2207.09837.
92. Stephanov, M.A. QCD phase diagram: An Overview. *PoS* **2006**, *LAT2006*, 024. [CrossRef]
93. Fukushima, K.; Hatsuda, T. The phase diagram of dense QCD. *Rept. Prog. Phys.* **2011**, *74*, 014001. [CrossRef]
94. Stephanov, M.A. On the sign of kurtosis near the QCD critical point. *Phys. Rev. Lett.* **2011**, *107*, 052301. [CrossRef] [PubMed]
95. Stephanov, M.; Yin, Y. Hydrodynamics with parametric slowing down and fluctuations near the critical point. *Phys. Rev. D* **2018**, *98*, 036006. [CrossRef]
96. Bazavov, A.; Ding, H.T.; Hegde, P.; Kaczmarek, O.; Karsch, F.; Laermann, E.; Maezawa, Y.; Mukherjee, S.; Ohno, H.; Wagner, M.; et al. The QCD Equation of State to $\mathcal{O}(\mu_B^6)$ from Lattice QCD. *Phys. Rev. D* **2017**, *95*, 054504. [CrossRef]
97. Fu, W.J.; Pawlowski, J.M.; Rennecke, F. QCD phase structure at finite temperature and density. *Phys. Rev. D* **2020**, *101*, 054032. [CrossRef]
98. Fu, W.j.; Luo, X.; Pawlowski, J.M.; Rennecke, F.; Wen, R.; Yin, S. Hyper-order baryon number fluctuations at finite temperature and density. *Phys. Rev. D* **2021**, *104*, 094047. [CrossRef]
99. Bzdak, A.; Esumi, S.; Koch, V.; Liao, J.; Stephanov, M.; Xu, N. Mapping the Phases of Quantum Chromodynamics with Beam Energy Scan. *Phys. Rept.* **2020**, *853*, 1–87. [CrossRef]
100. Bazavov, A.; Bollweg, D.; Ding, H.-T.; Enns, P.; Goswami, J.; Hegde, P.; Kaczmarek, O.; Karsch, F.; Larsen, R.; Mukherjee, S.; et al. Skewness, kurtosis, and the fifth and sixth order cumulants of net baryon-number distributions from lattice QCD confront high-statistics STAR data. *Phys. Rev. D* **2020**, *101*, 074502. [CrossRef]
101. Luo, X.; Wang, Q.; Xu, N.; Zhuang, P. (Eds.) *Properties of QCD Matter at High Baryon Density*; Springer: Berlin/Heidelberg, Germany, 2022. [CrossRef]
102. Gavai, R.V.; Gupta, S. The Critical end point of QCD. *Phys. Rev. D* **2005**, *71*, 114014. [CrossRef]
103. Cheng, M.; Hegde, P.; Jung, C.; Karsch, F.; Kaczmarek, O.; Laermann, E.; Mawhinney, R.D.; Miao, C.; Petreczky, P.; Schmidt, C.; et al. Baryon Number, Strangeness and Electric Charge Fluctuations in QCD at High Temperature. *Phys. Rev. D* **2009**, *79*, 074505. [CrossRef]
104. Gavai, R.V.; Gupta, S. Lattice QCD predictions for shapes of event distributions along the freezeout curve in heavy-ion collisions. *Phys. Lett. B* **2011**, *696*, 459–463. [CrossRef]
105. Bazavov, A.; Ding, H.-T.; Hegde, P.; Kaczmarek, O.; Karsch, F.; Laermann, E.; Mukherjee, S.; Petreczky, P.; Schmidt, C.; Smith, D.; et al. Freeze-out Conditions in Heavy Ion Collisions from QCD Thermodynamics. *Phys. Rev. Lett.* **2012**, *109*, 192302. [CrossRef]
106. Borsanyi, S.; Fodor, Z.; Katz, S.; Krieg, S.; Ratti, C.; Szabo, K.K. Freeze-out parameters from electric charge and baryon number fluctuations: Is there consistency? *Phys. Rev. Lett.* **2014**, *113*, 052301. [CrossRef] [PubMed]
107. Bellwied, R.; Borsanyi, S.; Fodor, Z.; Katz, S.D.; Ratti, C. Is there a flavor hierarchy in the deconfinement transition of QCD? *Phys. Rev. Lett.* **2013**, *111*, 202302. [CrossRef]
108. Adamczewski-Musch, J.; Arnold, O.; Behnke, C.; Belounnas, A.; Belyaev, A.; Berger-Chen, J.C.; Blanco, A.; Blume, C.; Böhmer, M.; Bordalo, P.; et al. Proton-number fluctuations in $\sqrt{s_{NN}} = 2.4$ GeV Au + Au collisions studied with the High-Acceptance DiElectron Spectrometer (HADES). *Phys. Rev. C* **2020**, *102*, 024914. [CrossRef]
109. Bass, S.; Belkacem, M.; Bleicher, M.; Brandstetter, M.; Bravina, L.; Ernst, C.; Gerland, L.; Hofmann, M.; Konopka, J.; Mao, G.; et al. Microscopic models for ultrarelativistic heavy ion collisions. *Prog. Part. Nucl. Phys.* **1998**, *41*, 255–369. [CrossRef]

110. Bleicher, M.; Zabrodin, E.; Spieles, C.; Bass, S.A.; Ernst, C.; Soff, S.; Bravina, L.; Belkacem, M.; Weber, H.; Stöcker, H.; et al. Relativistic hadron hadron collisions in the ultrarelativistic quantum molecular dynamics model. *J. Phys. G* **1999**, *25*, 1859–1896. [CrossRef]
111. Braun-Munzinger, P.; Friman, B.; Redlich, K.; Rustamov, A.; Stachel, J. Relativistic nuclear collisions: Establishing a non-critical baseline for fluctuation measurements. *Nucl. Phys. A* **2021**, *1008*, 122141. [CrossRef]
112. Vovchenko, V.; Koch, V.; Shen, C. Proton number cumulants and correlation functions in Au-Au collisions at sNN = 7.7–200 GeV from hydrodynamics. *Phys. Rev. C* **2022**, *105*, 014904. [CrossRef]
113. Abdallah, M.; Xu, Z. Higher-Order Cumulants and Correlation Functions of Proton Multiplicity Distributions in $\sqrt{s_{NN}}$ = 3 GeV Au+Au Collisions at the STAR Experiment. *arXiv* **2022**, arXiv:2209.11940.
114. Almaalol, D.; Hippert, M.; Noronha-Hostler, J.; Noronha, J.; Speranza, E.; Basar, G.; Bass, S.; Cebra, D.; Dexheimer, V.; Shen, C.; et al. QCD Phase Structure and Interactions at High Baryon Density: Completion of BES Physics Program with CBM at FAIR. *arXiv* **2022**, arXiv:2209.05009.
115. Sorensen, A.; Oliinychenko, D.; Koch, V.; McLerran, L. Speed of Sound and Baryon Cumulants in Heavy-Ion Collisions. *Phys. Rev. Lett.* **2021**, *127*, 042303. [CrossRef] [PubMed]
116. Gupta, S.; Mallick, D.; Mishra, D.K.; Mohanty, B.; Xu, N. Limits of thermalization in relativistic heavy ion collisions. *Phys. Lett. B* **2022**, *829*, 137021. [CrossRef]
117. Andronic, A.; Braun-Munzinger, P.; Redlich, K.; Stachel, J. Decoding the phase structure of QCD via particle production at high energy. *Nature* **2018**, *561*, 321–330. [CrossRef]
118. Gupta, S.; Luo, X.; Mohanty, B.; Ritter, H.G.; Xu, N. Scale for the Phase Diagram of Quantum Chromodynamics. *Science* **2011**, *332*, 1525–1528. [CrossRef]
119. Bzdak, A.; Koch, V.; Skokov, V. Baryon number conservation and the cumulants of the net proton distribution. *Phys. Rev. C* **2013**, *87*, 014901. [CrossRef]
120. The STAR Collaboration. Studying the Phase Diagram of QCD Matter at RHIC—A STAR Whitepaper for BES-II. 2014. Available online: https://drupal.star.bnl.gov/STAR/starnotes/public/sn0598 (accessed on 1 February 2023).
121. Rafelski, J.; Muller, B. Strangeness Production in the Quark - Gluon Plasma. *Phys. Rev. Lett.* **1982**, *48*, 1066; Erratum in *Phys. Rev. Lett.* **1986**, *56*, 2334. [CrossRef]
122. Koch, P.; Müller, B.; Rafelski, J. Strangeness in relativistic heavy ion collisions. *Phys. Rep.* **1986**, *142*, 167–262. [CrossRef]
123. Adams, J.; Adler, C.; Aggarwal, M.M.; Ahammed, Z.; Amonett, J.; Anderson, B.D.; Anderson, M.; Arkhipkin, D.; Averichev, G.S.; Badyal, S.K.; et al. Particle-Type Dependence of Azimuthal Anisotropy and Nuclear Modification of Particle Production in Au+Au Collisions at $\sqrt{s_{NN}} = 200$ GeV. *Phys. Rev. Lett.* **2004**, *92*, 052302. [CrossRef] [PubMed]
124. Adams, J.; Aggarwal, M.M.; Ahammed, Z.; Amonett, J.; Anderson, B.D.; Arkhipkin, D.; Averichev, G.S.; Badyal, S.K.; Bai, Y.; Balewski, J.; et al. Multistrange Baryon Elliptic Flow in Au + Au Collisions at $\sqrt{s_{NN}} = 200$ GeV. *Phys. Rev. Lett.* **2005**, *95*, 122301. [CrossRef] [PubMed]
125. Chen, J.H.; Jin, F.; Gangadharan, D.; Cai, X.Z.; Huang, H.Z.; Ma, Y.G. Parton distributions at hadronization from bulk dense matter produced in Au + Au collisions at $\sqrt{s_{NN}} = 200$ GeV. *Phys. Rev. C* **2008**, *78*, 034907. [CrossRef]
126. Adamczyk, L.; Adkins, J.K.; Agakishiev, G.; Aggarwal, M.M.; Ahammed, Z.; Alekseev, I.; Aparin, A.; Arkhipkin, D.; Aschenauer, E.C.; Attri, A.; et al. Probing parton dynamics of QCD matter with Ω and ϕ production. *Phys. Rev. C* **2016**, *93*, 021903. [CrossRef]
127. Adam, J.; Adamczyk, L.; Adams, J.R.; Adkins, J.K.; Agakishiev, G.; Aggarwal, M.M.; Ahammed, Z.; Alekseev, I.; Anderson, D.M.; Aoyama, R.; et al. Strange hadron production in Au + Au collisions at $\sqrt{s_{NN}} = 7.7, 11.5, 19.6, 27$, and 39 GeV. *Phys. Rev. C* **2020**, *102*, 034909. [CrossRef]
128. Wu, X. Baryon Number Transport, Strangeness Conservation and Ω-hadron Correlations. In Proceedings of the 20th International Conference on Strangeness in Quark Matter, Busan, Republic of Korea, 13–17 June 2022.
129. Adamczyk, L.; Adkins, J.K.; Agakishiev, G.; Aggarwal, M.M.; Ahammed, Z.; Alekseev, I.; Alford, J.; Anson, C.D.; Aparin, A.; Arkhipkin, D.; et al. $\Lambda\Lambda$ Correlation Function in Au + Au Collisions at $\sqrt{s_{NN}} = 200$ GeV. *Phys. Rev. Lett.* **2015**, *114*, 022301. [CrossRef] [PubMed]
130. Iritani, T.; Aoki, S.; Doi, T.; Etminan, F.; Gongyo, S.; Hatsuda, T.; Ikeda, Y.; Inoue, T.; Ishii, N.; Miyamoto, T.; et al. $N\Omega$ dibaryon from lattice QCD near the physical point. *Phys. Lett. B* **2019**, *792*, 284–289. [CrossRef]
131. Adam, J.; Adamczyk, L.; Adams, J.; Adkins, J.; Agakishiev, G.; Aggarwal, M.; Ahammed, Z.; Ajitanand, N.; Alekseev, I.; Anderson, D.; et al. The Proton-Ω correlation function in Au+Au collisions at $\sqrt{s_{NN}}$ = 200 GeV. *Phys. Lett. B* **2019**, *790*, 490–497. [CrossRef]
132. Deb, S.; Rath, R.; Roy, A.; Sahoo, R. Unveiling the strong interaction among hadrons at the LHC. *Nature* **2020**, *588*, 232–238; Erratum in *Nature* **2021**, *590*, E13. [CrossRef]
133. Andronic, A.; Braun-Munzinger, P.; Stachel, J.; Stocker, H. Production of light nuclei, hypernuclei and their antiparticles in relativistic nuclear collisions. *Phys. Lett. B* **2011**, *697*, 203–207. [CrossRef]
134. Andronic, A.; Braun-Munzinger, P.; Stachel, J. Hadron production in central nucleus-nucleus collisions at chemical freeze-out. *Nucl. Phys. A* **2006**, *772*, 167–199. [CrossRef]
135. Steinheimer, J.; Gudima, K.; Botvina, A.; Mishustin, I.; Bleicher, M.; Stöcker, H. Hypernuclei, dibaryon and antinuclei production in high energy heavy ion collisions: Thermal production versus Coalescence. *Phys. Lett. B* **2012**, *714*, 85–91. [CrossRef]

136. Zhang, S.; Chen, J.; Crawford, H.; Keane, D.; Ma, Y.; Xu, Z. Searching for onset of deconfinement via hypernuclei and baryon-strangeness correlations. *Phys. Lett. B* **2010**, *684*, 224–227. [CrossRef]
137. Guo, Y.; Liao, J.; Wang, E.; Xing, H.; Zhang, H. Hyperon polarization from the vortical fluid in low-energy nuclear collisions. *Phys. Rev. C* **2021**, *104*, L041902. [CrossRef]
138. Ivanov, Y.B. Global Λ polarization in moderately relativistic nuclear collisions. *Phys. Rev. C* **2021**, *103*, L031903. [CrossRef]
139. Adam, J.; Adamová, D.; Aggarwal, M.; Rinella, G.A.; Agnello, M.; Agrawal, N.; Ahammed, Z.; Ahn, S.; Aimo, I.; Aiola, S.; et al. ${}^{3}_{\Lambda}\mathrm{H}$ and ${}^{3}_{\Lambda}\overline{\mathrm{H}}$ production in Pb-Pb collisions at $\sqrt{s_{NN}} = 2.76$ TeV. *Phys. Lett. B* **2016**, *754*, 360–372. [CrossRef]
140. Acharya, S.; Adamová, D.; Adhya, S.; Adler, A.; Adolfsson, J.; Aggarwal, M.; Rinella, G.A.; Agnello, M.; Agrawal, N.; Ahammed, Z.; et al. ${}^{3}_{\Lambda}\mathrm{H}$ and ${}^{3}_{\Lambda}\overline{\mathrm{H}}$ lifetime measurement in Pb-Pb collisions at $\sqrt{s_{NN}} = 5.02$ TeV via two-body decay. *Phys. Lett. B* **2019**, *797*, 134905. [CrossRef]
141. Gläßel, S.; Kireyeu, V.; Voronyuk, V.; Aichelin, J.; Blume, C.; Bratkovskaya, E.; Coci, G.; Kolesnikov, V.; Winn, M. Cluster and hypercluster production in relativistic heavy-ion collisions within the parton-hadron-quantum-molecular-dynamics approach. *Phys. Rev. C* **2022**, *105*, 014908. [CrossRef]
142. Friedman, E.; Gal, A. ΛNN content of Λ-nucleus potential. *EPJ Web Conf.* **2022**, *271*, 06002. [CrossRef]
143. Adams, J.; Aggarwal, M.; Ahammed, Z.; Amonett, J.; Anderson, B.; Arkhipkin, D.; Averichev, G.; Badyal, S.; Bai, Y.; Balewski, J.; et al. Experimental and theoretical challenges in the search for the quark gluon plasma: The STAR Collaboration's critical assessment of the evidence from RHIC collisions. *Nucl. Phys. A* **2005**, *757*, 102–183. [CrossRef]
144. Luo, X.; Xu, N. Search for the QCD Critical Point with Fluctuations of Conserved Quantities in Relativistic Heavy-Ion Collisions at RHIC: An Overview. *Nucl. Sci. Tech.* **2017**, *28*, 112. [CrossRef]
145. Cleymans, J.; Oeschler, H.; Redlich, K.; Wheaton, S. Comparison of chemical freeze-out criteria in heavy-ion collisions. *Phys. Rev. C* **2006**, *73*, 034905. [CrossRef]

symmetry

MDPI

Article

Production of Strange and Charm Hadrons in Pb+Pb Collisions at $\sqrt{s_{NN}} = 5.02$ TeV †

Wen-Bin Chang [1,2], Rui-Qin Wang [1], Jun Song [3], Feng-Lan Shao [1,*], Qun Wang [4,*] and Zuo-Tang Liang [5,*]

1 School of Physics and Physical Engineering, Qufu Normal University, Qufu 273165, China
2 Institute of Particle Physics and Key Laboratory of Quark and Lepton Physics (MOS), Central China Normal University, Wuhan 430079, China
3 School of Physical Science and Intelligent Engineering, Jining University, Qufu 273155, China
4 Department of Modern Physics, University of Science and Technology of China, Hefei 230026, China
5 Key Laboratory of Particle Physics and Particle Irradiation (MOE), Institute of Frontier and Interdisciplinary Science, Shandong University, Qingdao 266237, China
* Correspondence: shaofl@mail.sdu.edu.cn (F.-L.S.); qunwang@ustc.edu.cn (Q.W.); liang@sdu.edu.cn (Z.-T.L.)
† This article is dedicated to the memory of Qu-bing Xie.

Abstract: Using a quark combination model with the equal-velocity combination approximation, we study the production of hadrons with strangeness and charm flavor quantum numbers in Pb+Pb collisions at $\sqrt{s_{NN}} = 5.02$ TeV. We present analytical expressions and numerical results for these hadrons' transverse momentum spectra and yield ratios. Our numerical results agree well with the experimental data available. The features of strange and charm hadron production in the quark–gluon plasma at the early stage of heavy ion collisions are also discussed.

Keywords: heavy-ion collisions; heavy-flavor production; hadronization

PACS: 27.75.-q; 25.75.Ag; 25.75.Dw; 25.75.Gz

1. Introduction

It is well-known that the hadronic matter is expected to undergo a transition to the quark–gluon plasma (QGP), a strongly coupled state of matter, at high temperatures or baryon densities [1–6]. The search for the QGP and the study of its properties have long been the goals of high-energy heavy ion collisions [7–10]. In heavy ion collisions, strange and heavy-flavor quarks are newly produced (or excited from the vacuum) and most of them are present in the whole stage of the QGP evolution. They interact strongly with the constituents of the QGP medium or they are a part of the QGP. Therefore, strange and heavy-flavor hadrons are usually regarded as special probes to the hadronization mechanism and properties of the QGP [11–22].

The relativistic heavy ion collider (RHIC) and the large hadron collider (LHC) have accumulated abundant experimental data on strange and charm hadrons [23–36]. These data show a number of features in the production of strangeness [33,34,37–40] and baryons [32,35,41–44]. Many efforts have been made to understand hadron production mechanisms in theory and phenomenology [45–62]. Hydrodynamic and thermal models [45–49] are commonly used to describe the production of strange hadrons. For the production of heavy-flavor hadrons, some transport models are popular (see, e.g., reference [63] and references therein). In particular, the coalescence or recombination models also provide good descriptions of hadron production especially at low and intermediate transverse momenta [50–56,64].

Based on Qu-Bing Xie's works in e^+e^- and pp collisions in early years [65–70], we developed a quark combination model (QCM) for hadronization and it works well in explaining yields, rapidity distributions, and transverse momentum spectra for the identified hadrons in high-energy heavy ion collisions at various energies ranging from

RHIC to LHC [53,62,71,72]. Recently, inspired by the property of constituent quark number scaling for transverse momentum spectra of strange hadrons in p+Pb collisions at LHC energy [73], we proposed a simplified version of the quark combination model by incorporating the equal-velocity combination (EVC) to replace the near rapidity combination in the original model. Many properties of hadron production can be analytically derived and some of them have been tested by experimental data in high energy pp, pA, and AA collisions [74–78]. Furthermore, our studies show that the EVC of charm and light quarks can explain the transverse momentum spectra of single-charm hadrons at low and intermediate transverse momenta [75,77–80]. In particular, the model prediction of the Λ_c^+/D^0 ratio was verified by the latest measurements of the ALICE collaboration [35,81,82].

Recently, the ALICE collaboration published precise measurements of strange and charm hadrons, especially D mesons and Λ_c^+ baryons, in Pb+Pb collisions at LHC [31,83–86]. In this paper, we apply the QCM with EVC to study the production of strange and charm hadrons simultaneously at low and intermediate transverse momenta in Pb+Pb collisions at $\sqrt{s_{NN}} = 5.02$ TeV. We will present analytical and numerical results for the p_T dependence of production ratios between different strange and charm hadrons. We will compare our results with the experimental data available and make predictions for other types of hadrons.

The rest of the paper is organized as follows. In Section 2, we introduce a general phase-space structure of the QCM in heavy ion collisions as well as the idea and formula of the QCM in momentum space based on EVC. In Sections 3 and 4, we apply the QCM to calculate spectra of various strange and charm hadrons in Pb+Pb collisions at $\sqrt{s_{NN}} = 5.02$ TeV and compare them with data. The final section is a summary of the main results and conclusions.

2. The Quark Combination Model

The QCM developed by the Shandong group led by Qu-Bing Xie [65–70] is a kind of exclusive or statistical hadronization model with constituent quarks as building blocks. A quark combination rule (QCR) can be derived for quarks and antiquarks in the neighborhood of the longitudinal phase space (momentum rapidity) to combine into baryons and mesons [65–68,87]. The QCM based on the QCR has successfully explained experimental data on hadron production in e^+e^- and pp collisions [65–70,88] as well as in heavy ion collisions [53,71,89]. A modern version of QCM with spin degrees of freedom in terms of Wigner functions has been developed by some of us and applied to spin polarization of hadrons in heavy ion collisions [90–93].

In this section, we introduce the general phase-space structure of the QCM in heavy ion collisions as well as its simplified version in momentum space to describe momentum spectra of strange and charm hadrons.

2.1. General Phase Space Structure of QCM in Heavy Ion Collisions

In the quantum kinetic theory, the formation of a composite particle through the coalescence or combination process of its constituent particles $q_1q_2\cdots q_n \to H$ can be described by the collision term incorporating the matrix element squared of the process and momentum integrals. In the case we are considering, the composite particle H can be a meson or a baryon, so the constituent particles $q_1q_2\cdots q_n$ are a quark and an antiquark for the meson, and are three quarks or three antiquarks for the baryon or antibaryon, respectively. In heavy ion collisions, the coalescence process takes place in a space-time region, i.e., the freeze-out hypersurface defined by the proper time τ_0. The momentum distribution of the hadron (meson or baryon) reads

$$f_H(\mathbf{p}) \sim \int d\sigma^\mu p_\mu \int \prod_{i=1}^{n} \frac{d^3\mathbf{p}_i}{(2\pi)^3 2E_i}(2\pi)^4\delta(p_1+p_2+\cdots+p_n-p) \times |M(q_1q_2\cdots q_n \to H)|^2 f_1(x,p_1)f_2(x,p_2)\cdots f_n(x,p_n), \quad (1)$$

where $p = (E_p, \mathbf{p})$ is the hadron's on-shell momentum, $p_i = (E_i, \mathbf{p}_i)$ is the on-shell momentum of the constituent particle q_i with its momentum distribution $f_i(x, p_i)$ at the space-time point x on the freeze-out hypersurface, M is the invariant amplitude of the coalescence process containing the hadron's wave function, and $d\sigma^\mu(x)$ is the surface element pointing to the normal direction of the freeze-out hypersurface at x. The momentum distribution can be decomposed into the thermal part and non-thermal part,

$$f_i(x, p_i) = f_i^{\text{th}}(\beta u \cdot p_i) + f_i^{\text{nth}}(p_i), \tag{2}$$

where the thermal part f_i^{th} depends on $\beta u \cdot p_i$ with $\beta(x) = 1/T(x)$ being the inverse temperature and $u^\mu(x)$ being the flow velocity both of which are functions of x on the freeze-out hypersurface, and the non-thermal part f_i^{nth} depends only on momentum and is independent of the space-time coordinate. We can express the space-time point on the freeze-out hypersurface in terms of the proper time τ and space-time rapidity η as

$$x^\mu = (\tau \cosh \eta, \mathbf{x}_T, \tau \sinh \eta), \tag{3}$$

and also the hadron's on-shell momentum in terms of transverse momentum $\mathbf{p}_T$ and rapidity Y as

$$p^\mu = (m_T \cosh Y, \mathbf{p}_T, m_T \sinh Y), \tag{4}$$

where $m_T = \sqrt{m^2 + p_T^2}$ is the transverse mass. Then the freeze-out hypersurface element can be expressed as

$$d\sigma^\mu = \tau d\eta d^2 x_T \frac{\partial x^\mu}{\partial \tau} = \tau d\eta d^2 x_T (\cosh \eta, 0, 0, \sinh \eta), \tag{5}$$

so its contraction with the hadron's momentum reads

$$d\sigma^\mu p_\mu = \tau d\eta d^2 x_T m_T \cosh(\eta - Y). \tag{6}$$

We can express the flow velocity with Bjorken's boost invariance in the longitudinal direction with $\eta = \eta_{\text{flow}}$,

$$\begin{aligned} u^\mu(x) =& [\cosh \eta \cosh \rho(x_T, \phi_s), \sinh \rho(x_T, \phi_s) \cos \phi_b, \\ & \sinh \rho(x_T, \phi_s) \sin \phi_b, \sinh \eta \cosh \rho(x_T, \phi_s)], \end{aligned} \tag{7}$$

where $\rho(x_T, \phi_s)$ is the transverse flow rapidity [94,95] as a function of cylindrical coordinates in the transverse plane $x_T = |\mathbf{x}_T|$ and ϕ_s, and ϕ_b is the boost angle in the transverse plane which can simply be taken as ϕ_s in approximation. The elliptic flow can be implemented by [95]

$$\rho(x_T, \phi_s) = \frac{x_T}{R}[\rho_0 + \rho_2 \cos(2\phi_s)], \tag{8}$$

where R is the transverse size of the fireball, and ρ_2 is linked to the elliptic flow coefficient v_2.

2.2. QCM in Momentum Space with Equal-Velocity Combination

For the purpose of this paper, we will introduce a simplified version of the QCM in momentum space with an equal-velocity combination for hadron production. This corresponds to (a) the quark distributions are homogeneous in space-time and depend only on momentum and (b) the role of the matrix element squared is taken by the EVC. This version of QCM is an approximation to the rigorous one in Section 2.1.

We consider a color-neutral system of $N_q = \sum_i N_{q_i}$ quarks and $N_{\bar{q}} = \sum_i N_{\bar{q}_i}$ antiquarks where $q_i = u, d, s, c$ and $\bar{q}_i = \bar{u}, \bar{d}, \bar{s}, \bar{c}$ denote the quark and antiquark flavors, respectively. The momentum distributions $f_{M_j}(p) \equiv f_{M_j}(p; N_q, N_{\bar{q}})$ and $f_{B_j}(p) \equiv f_{B_j}(p; N_q, N_{\bar{q}})$ for the directly produced meson M_j and baryon B_j by combining a pair of quark–antiquarks and three quarks, respectively, can be schematically expressed as

$$f_{M_j}(p) = \sum_{\bar{q}_1 q_2} \int dp_1 dp_2 N_{\bar{q}_1 q_2} f^{(n)}_{\bar{q}_1 q_2}(p_1, p_2) \mathcal{R}_{M_j,\bar{q}_1 q_2}(p; p_1, p_2), \tag{9}$$

$$f_{B_j}(p) = \sum_{q_1 q_2 q_3} \int dp_1 dp_2 dp_3 N_{q_1 q_2 q_3} f^{(n)}_{q_1 q_2 q_3}(p_1, p_2, p_3) \mathcal{R}_{B_j,q_1 q_2 q_3}(p; p_1, p_2, p_3), \tag{10}$$

where $f^{(n)}_{\bar{q}_1 q_2}$ and $f^{(n)}_{q_1 q_2 q_3}$ are normalized joint momentum distributions; $N_{\bar{q}_1 q_2}$ and $N_{q_1 q_2 q_3}$ are the number of $\bar{q}_1 q_2$ pairs and that of $q_1 q_2 q_3$ clusters in the system; $\mathcal{R}_{M_j,\bar{q}_1 q_2}$ and $\mathcal{R}_{B_j,q_1 q_2 q_3}$ are combination kernel functions that stand for the probability density for a $\bar{q}_1 q_2$ pair with momenta p_1 and p_2 to combine into a meson M_j of momentum p and that for a $q_1 q_2 q_3$ cluster with p_1, p_2, and p_3 to combine into a baryon B_j of momentum p, respectively.

Just as derived in Refs. [75,76], the combination kernel functions in the EVC can be written as

$$\mathcal{R}_{M_j,\bar{q}_1 q_2}(p; p_1, p_2) = C_{M_j} \mathcal{R}^{(f)}_{\bar{q}_1 q_2} \mathcal{A}_{M,\bar{q}_1 q_2} \delta(p_1 - x^{q_1}_{q_1 q_2} p) \delta(p_2 - x^{q_2}_{q_1 q_2} p), \tag{11}$$

$$\begin{aligned} \mathcal{R}_{B_j,q_1 q_2 q_3}(p; p_1, p_2, p_3) = {} & C_{B_j} \mathcal{R}^{(f)}_{q_1 q_2 q_3} \mathcal{A}_{B,q_1 q_2 q_3} \delta(p_1 - x^{q_1}_{q_1 q_2 q_3} p) \\ & \times \delta(p_2 - x^{q_2}_{q_1 q_2 q_3} p) \delta(p_3 - x^{q_3}_{q_1 q_2 q_3} p), \end{aligned} \tag{12}$$

where δ-functions guarantee the momentum conservation in the EVC and $x^{q_i}_{q_1 q_2} = m_{q_i}/(m_{q_1} + m_{q_2})$ and $x^{q_i}_{q_1 q_2 q_3} = m_{q_i}/(m_{q_1} + m_{q_2} + m_{q_3})$ are the momentum fraction of the produced hadron for q_i. We note that the mass fraction is the same as the momentum fraction in the EVC. Masses of up, down, strange, and charm quarks are taken to be $m_u = m_d = 0.3$ GeV, $m_s = 0.5$ GeV and $m_c = 1.5$ GeV, respectively.

The factor C_{M_j} is the probability for M to be M_j if the quark content of M is the same as M_j and similar for C_{B_j}. In this paper, we only consider hadrons in the ground state, namely mesons with $J^P = 0^-$ and 1^- and baryons with $J^P = (1/2)^+$ and $(3/2)^+$. In this case, C_{M_j} is the same for all hadrons in the same multiplet (with the same J^P) and determined by the production ratio of vector to pseudo-scalar mesons $R_{V/P}$, so is it for C_{B_j} which is determined by the production ratio of $J^P = (1/2)^+$ to $J^P = (3/2)^+$ baryons $R_{O/D}$ with the same flavor content.

The factors $\mathcal{R}^{(f)}_{\bar{q}_1 q_2}$ and $\mathcal{R}^{(f)}_{q_1 q_2 q_3}$ contain Kronecker δ's to guarantee the quark flavor conservation, e.g., if M_j is a D-meson with constituent quark content $\bar{q}c$, $\mathcal{R}^{(f)}_{\bar{q}_1 q_2} = \delta_{q_1,q}\delta_{q_2,c}$. If B_j is a single-charm baryon with the quark content udc, $\mathcal{R}^{(f)}_{q_1 q_2 q_3} = N_{\rm sym}\delta_{q_1,u}\delta_{q_2,d}\delta_{q_3,c}$, where $N_{\rm sym} = 1, 3, 6$ is a symmetry factor to account for the number of different permutations of three quarks for (a) three identical flavors, (b) two identical flavors, and (c) all three distinct flavors, respectively.

The factor $\mathcal{A}_{M,\bar{q}_1 q_2}$ is the probability for a quark q_2 to capture a specific antiquark $\bar{q}_1$ to form a meson in the quark–antiquark system; it should be inversely proportional to $N_q + N_{\bar{q}}$. Similarly, $\mathcal{A}_{B,q_1 q_2 q_3}$ should be inversely proportional to $(N_q + N_{\bar{q}})^2$. Both $\mathcal{A}_{M,\bar{q}_1 q_2}$ and $\mathcal{A}_{B,q_1 q_2 q_3}$ are determined by the unitarity and the competition mechanism of meson-baryon production. Note that for light-quark systems produced in e^+e^- and pp collisions, $\mathcal{A}_{M,\bar{q}_1 q_2}$ and $\mathcal{A}_{B,q_1 q_2 q_3}$ correspond to combination weights of mesons and baryons that follow the QCR [65,66,87].

Putting all these factors together, for charm hadrons we are considering, Equations (9) and (10) become

$$f_{M_j}(p) = \frac{N_c N_{\bar{q}_1}}{N_q + N_{\bar{q}}} \mathcal{A}_M C_{M_j} f^{(n)}_{\bar{q}_1 c}(x^{q_1}_{q_1 c} p, x^{c}_{q_1 c} p), \tag{13}$$

$$f_{B_j}(p) = \frac{N_c N_{q_1} N_{q_2}}{(N_q + N_{\bar{q}})^2} \mathcal{A}_B C_{B_j} N_{\rm sym} f^{(n)}_{q_1 q_2 c}(x^{q_1}_{q_1 q_2 c} p, x^{q_2}_{q_1 q_2 c} p, x^{c}_{q_1 q_2 c} p), \tag{14}$$

where $\mathcal{A}_M$ and $\mathcal{A}_B$ are two global coefficients that can be determined by quark number conservation in the combination process and the baryon-to-meson production ratio N_B/N_M. We are considering a quark–antiquark system in the mid-rapidity region at very high collision energies, so that net baryon number and net quark flavor are negligible, i.e., $N_{\bar{q}_i} \approx N_{q_i}$ for $i = u, d, s, c$. Moreover, we assume that the number of strange quarks is suppressed by a factor λ_s (strangeness suppression factor) relative to that of up and down quarks, so we have $N_u : N_d : N_s = 1 : 1 : \lambda_s$.

If we neglect correlations in the joint momentum distributions among different momenta, we have factorization forms for the joint momentum distributions,

$$f^{(n)}_{q_1 q_2}(p_1, p_2) = f^{(n)}_{q_1}(p_1) f^{(n)}_{q_2}(p_2), \quad (15)$$

$$f^{(n)}_{q_1 q_2 q_3}(p_1, p_2, p_3) = f^{(n)}_{q_1}(p_1) f^{(n)}_{q_2}(p_2) f^{(n)}_{q_3}(p_3). \quad (16)$$

We will use the above factorization forms in Equations (13) and (14) in our numerical calculation for single-charm hadrons. By using Equations (13) and (14) with Equations (15) and (16), we are able to calculate momentum spectra and yields for different hadrons.

Including strong and electromagnetic decay contributions from short-lived resonances [96], we can obtain the momentum spectra of final state hadrons and make comparison with experimental data. For charm hadrons, we make an approximation that the momentum of the daughter charm hadron is almost equal to that of the mother charm hadron. With this approximation and the production ratio of the vector to the pseudo-scalar meson being set to 1.5 [28,80], we obtain (for the final state D mesons):

$$f^{(\mathrm{fin})}_{D^0}(p) \approx 3.516 f_{D^0}(p), \quad (17)$$

$$f^{(\mathrm{fin})}_{D^+}(p) \approx 1.485 f_{D^+}(p), \quad (18)$$

$$f^{(\mathrm{fin})}_{D_s^+}(p) \approx 2.5 f_{D_s^+}(p). \quad (19)$$

Similarly, we can set the production ratio of $J^P = (1/2)^+$ to the $J^P = (3/2)^+$ single-charm baryon to 2 [80] and obtain,

$$f^{(\mathrm{fin})}_{\Lambda_c^+}(p) \approx 5 f_{\Lambda_c^+}(p), \quad (20)$$

$$f^{(\mathrm{fin})}_{\Sigma_c^0}(p) \approx f_{\Sigma_c^0}(p), \quad (21)$$

$$f^{(\mathrm{fin})}_{\Sigma_c^+}(p) \approx f_{\Sigma_c^+}(p), \quad (22)$$

$$f^{(\mathrm{fin})}_{\Sigma_c^{++}}(p) \approx f_{\Sigma_c^{++}}(p), \quad (23)$$

$$f^{(\mathrm{fin})}_{\Xi_c^0}(p) \approx 2.5 f_{\Xi_c^0}(p), \quad (24)$$

$$f^{(\mathrm{fin})}_{\Xi_c^+}(p) \approx 2.5 f_{\Xi_c^+}(p), \quad (25)$$

$$f^{(\mathrm{fin})}_{\Omega_c^0}(p) \approx 1.5 f_{\Omega_c^0}(p). \quad (26)$$

These analytical results can be used to obtain the p_T spectra of charm hadrons. For final state strange hadrons, there are no such analytical results, only numerical ones.

3. Transverse Momentum Spectra and Baryon-to-Meson Ratio for Strange Hadrons

In this section, we apply the QCM introduced in Section 2 to study the production of strange hadrons in Pb+Pb collisions at $\sqrt{s_{NN}} = 5.02$ TeV. We first calculate the p_T spectra of strange mesons and baryons. Then we calculate the baryon-to-meson ratio Λ/K_S^0 as a function of p_T in different types of centralities.

3.1. Transverse Momentum Spectra of Strange Hadrons

The inputs of the model are p_T spectra of quarks and antiquarks. In this paper, we adopt the isospin symmetry and neglect the net quark numbers ($N_{q_i} \approx N_{\bar{q}_i}$ for $i = u, d, s, c$) in the mid-rapidity region at LHC energy, so we have only two inputs $f_d(p_T) = f_u(p_T)$ and $f_s(p_T)$, which can be fixed by fitting the experimental data on the p_T spectra of ϕ mesons and Λ baryons [40,97,98]. The extracted results for the normalized p_T spectra of quarks in central 0–5% to peripheral 70–80% Pb+Pb collisions at $\sqrt{s_{NN}} = 5.02$ TeV are shown in Figure 1. The rapidity densities of d and s quarks are listed in Table 1.

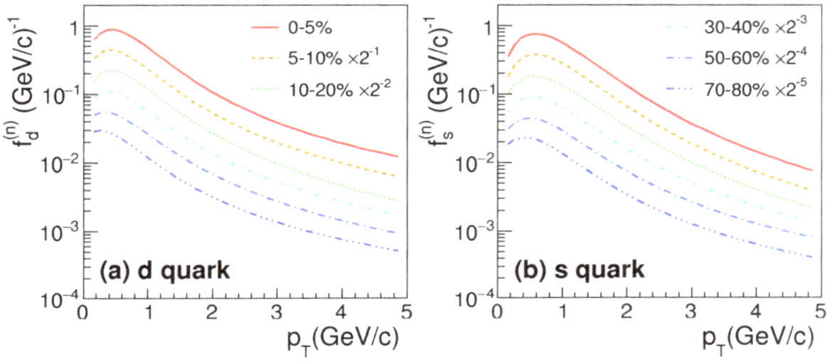

Figure 1. The normalized p_T distributions of (**a**) d quark and (**b**) s quark in different centralities in Pb+Pb collisions at $\sqrt{s_{NN}} = 5.02$ TeV.

Table 1. Rapidity densities of d and s quarks in different centralities in Pb+Pb collisions at $\sqrt{s_{NN}} = 5.02$ TeV.

Centrality	dN_d/dy	dN_s/dy
0–5%	840	370
5–10%	686	302
10–20%	516	227
30–40%	267	115
50–60%	97	40
70–80%	27	10

In Figure 2, we show the results for the p_T spectra of K_S^0 and Λ in 0–5%, 5–10%, 10–20%, 30–40%, 50–60%, 70–80% centralities, and those of ϕ, Ξ^- and Ω^- in 0–10%, 10–20%, 30–40%, 50–60%, 70–80% centralities. The QCM results are displayed in lines and experimental data [97,98] are displayed in open symbols. From Figure 2, we see that our QCM results for strange mesons and baryons agree with the experimental data very well. Such a good agreement provides a piece of evidence for the EVC mechanism in describing strange hadron production in Pb+Pb collisions at the LHC energy.

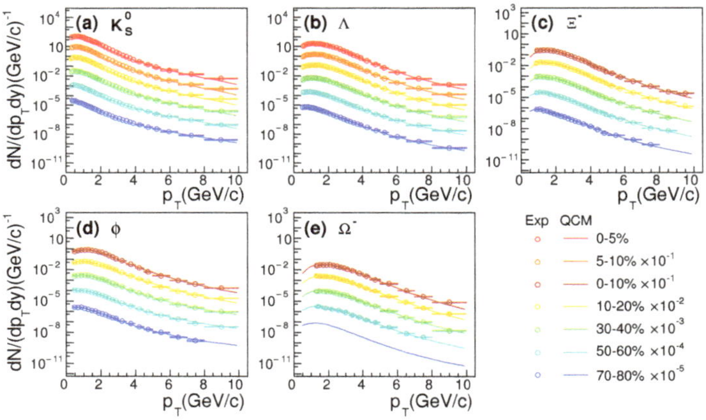

Figure 2. The p_T spectra of (**a**) K_S^0, (**b**) Λ, (**c**) Ξ^-, (**d**) ϕ and (**e**) Ω^- in different centralities in Pb+Pb collisions at $\sqrt{s_{NN}} = 5.02$ TeV. Open symbols are experimental data [97,98]. Lines for Λ and ϕ are fitting results that are used to fix p_T spectra of light-flavor quarks at hadronization in QCM. Lines for K_S^0, Ξ^- and Ω^- are predictions from QCM.

3.2. Baryon-To-Meson Ratio Λ/K_S^0

Figure 3 shows the multiplicity ratio Λ/K_S^0 as a function of p_T in five centrality ranges 0–5%, 10–20%, 30–40%, 50–60%, and 70–80%. Filled squares are experimental data [98], and lines are the QCM results. We see that Λ/K_S^0 exhibits an increase-peak-decrease behavior as a function of p_T in all centralities, which is regarded as a natural consequence of quark recombination [50–52,99,100]. We see that this feature can be well described by the QCM with EVC. The height of the peak increases from about 0.8 in the peripheral (70–80% centrality) to about 1.5 in the most central (0-5% centrality) collisions. The peak positions in the p_T slightly move to higher values from peripheral to central collisions due to stronger radial flows in more central collisions. The QCM with EVC gives a good description of the p_T dependence of Λ/K_S^0.

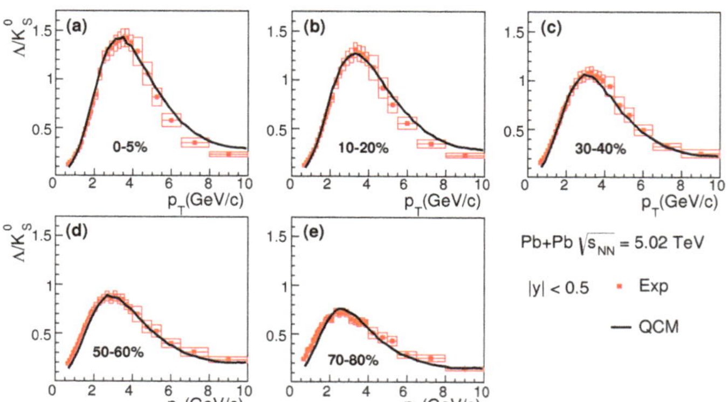

Figure 3. The p_T dependence of Λ/K_S^0 in (**a**) 0–5%, (**b**) 10–20%, (**c**) 30–40%, (**d**) 50–60%, (**e**) 70–80% centralities in Pb+Pb collisions at $\sqrt{s_{NN}} = 5.02$ TeV. Filled squares are experimental data [98] and lines are the QCM results.

4. Transverse Momentum Spectra, Yield Ratios, and Nuclear Modification Factor for Charm Hadrons

In this section, we apply the QCM with EVC to study the production of charm hadrons at midrapidity in Pb+Pb collisions at $\sqrt{s_{NN}} = 5.02$ TeV. We first calculate the p_T spectra of D mesons and single-charm baryons. Then we give the p_T dependence of yield ratios of different charm hadrons. Finally, we give the nuclear modification factor R_{AA} for charm hadrons as a function of p_T.

4.1. Transverse Momentum Spectra of Charm Mesons and Baryons

In the QCM, the only additional input is the normalized p_T distribution of the charm quarks, which we adopt a hybrid form based on the simulation of charm quarks propagating in the QGP medium in a Boltzmann transport approach [100,101]

$$f_c^{(n)}(p_T) = \frac{1}{N_{\rm norm}} p_T \left[\left(\frac{p_T}{p_{T0}}\right)^{\alpha_c} \exp\left(-\frac{\sqrt{p_T^2+m_c^2}}{T_c}\right) + \left(1.0 + \frac{\sqrt{p_T^2+m_c^2}-m_c}{\Gamma_c}\right)^{-\beta_c} \right]. \tag{27}$$

At small p_T, this parameterized form is very close to the thermal distribution, while at large p_T, it follows the power law which is a non-thermal distribution. Both the thermal and non-thermal distributions are smoothly connected through the above parameterization. Here, the normalization constant $N_{\rm norm}$ can be determined by the condition $\int_0^\infty dp_T f_c^{(n)}(p_T) = 1$. The parameters α_c, p_{T0}, T_c, Γ_c, and β_c are fitted using the data of D^0's p_T spectra [31,84] and are listed in Table 2. The shapes of $f_c^{(n)}(p_T)$ at different centralities are shown in Figure 4a. We see that there is a stronger suppression in more central collisions in the p_T range 4 GeV $< p_T <$ 10 GeV. Figure 4b shows $f_c^{(n)}(p_T)$ at different centralities normalized by 60–80% centrality, which has similar behavior to the nuclear modification factor R_{CP} of the D^0 meson measured in reference [31]. For the rapidity density of charm quarks dN_c/dy, we assume that it is proportional to the cross-section per rapidity in pp collisions as

$$\frac{dN_c}{dy} = \langle T_{AA} \rangle \frac{d\sigma_c^{\rm pp}}{dy}. \tag{28}$$

Here, $\langle T_{AA} \rangle$ is the average nuclear overlap function [31], $d\sigma_c^{pp}/dy$ is the p_T integrated cross-section of charm quarks in pp collisions which is about 1.0 mb at $\sqrt{s} = 5.02$ TeV [77]. The values of dN_c/dy at different centralities are listed in Table 2.

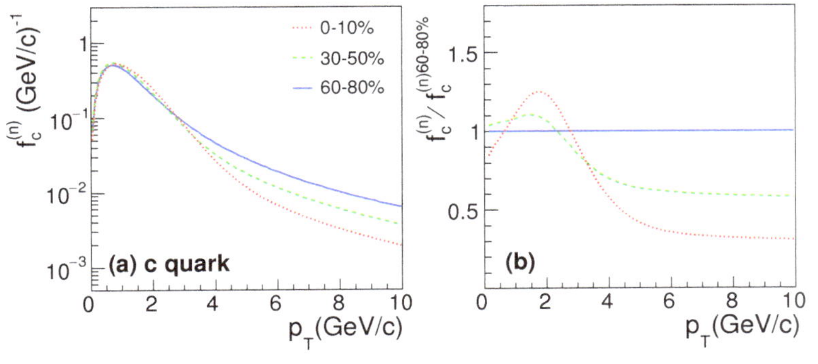

Figure 4. (**a**) Normalized p_T distribution of the charm quarks $f_c^{(n)}(p_T)$ at different centralities. (**b**) The distribution $f_c^{(n)}(p_T)$ at different centralities normalized by that at 60–80% centrality.

Table 2. The parameters in the normalized p_T distribution of the charm quarks at different centralities.

Centrality	0–10%	30–50%	60–80%
p_{T0} (GeV/c)	0.0051	0.10	0.63
α_c	0.5	1.0	1.5
T_c (GeV)	0.46	0.38	0.34
β_c	3.10	3.00	2.95
Γ_c (GeV)	0.6	0.6	0.7
dN_c/dy	23.07	3.90	0.417

In Figure 5, we present the results for the p_T spectra of D^0, D^+, D_s^+, and D^{*+} mesons at 0–10%, 30–50%, and 60–80% centralities. Open symbols are the experimental data [31,84,85] and lines are the calculated results. The results agree with the experimental data in the p_T range from 0.5 GeV/c up to 10 GeV/c.

Figure 6 shows the results for charm baryons at 0–10%, 30–50%, and 60–80% centralities. Open symbols [only in Figure 6a] are the data of Λ_c^+ [86] and lines are the results from the QCM, which are in good agreement. Predictions from QCM for other charm baryons Σ_c^0, Ξ_c^+, and Ω_c^0 are presented in Figure 6b–d, which can be tested in future experimental measurements.

The agreement between our results and experimental data for various D mesons and Λ_c^+ baryons indicates the validity of the EVC mechanism in the QCM in describing the charm hadron production. In this mechanism, the formation of the charm hadron is through the capture of light quarks in the medium by the charm quark with the same velocity.

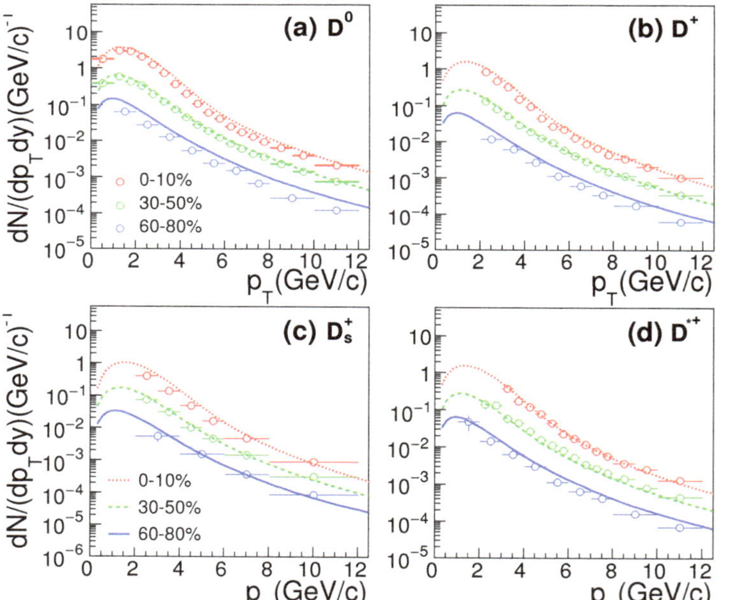

Figure 5. The p_T spectra of (**a**) D^0, (**b**) D^+, (**c**) D_s^+, and (**d**) D^{*+} mesons at different centralities. Open symbols are the experimental data [31,84,85]. Lines for D^0 are fitting results that are used to fix the shape parameters in the p_T spectrum of charm quarks at hadronization in Equation (27). Lines for D^+, D_s^+ and D^{*+} are predictions from QCM.

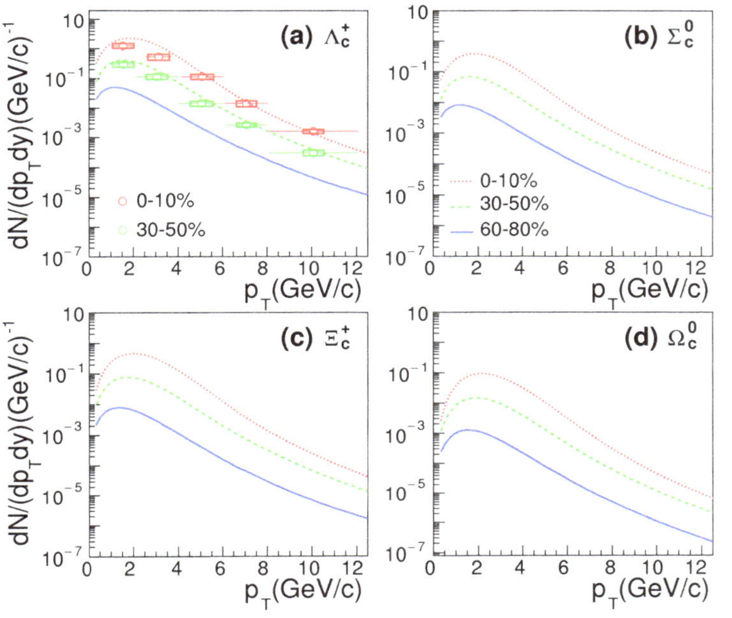

Figure 6. The p_T spectra of (**a**) Λ_c^+, (**b**) Σ_c^0, (**c**) Ξ_c^+, and (**d**) Ω_c^0 baryons at different centralities. Open symbols are the data of Λ_c^+ [86], and lines are the QCM results.

We also calculated p_T-integrated yield density dN/dy for charm hadrons at midrapidity and 0–10%, 30–50%, and 60–80% centralities as listed in Table 3. The experimental data are taken from Refs. [84,85]. The QCM results are slightly higher than data. This is because we only include single-charm hadrons in the calculation and assume all charm quarks go to single-charm hadrons. In fact, the hidden-charm J/Ψ, double-charm baryons, and even heavy-flavor multiquark states can be produced. If including these particles, our results in Table 3 will decrease slightly and the agreement with experimental data will be improved. No data are available for the yield densities of many charm hadrons; our QCM predictions can be tested by their future experimental measurements.

Table 3. The yield density dN/dy for charm hadrons at different centralities. The experimental data are from Refs. [84,85].

Hadron	0–10%		30–50%		60–80%	
	Data	QCM	Data	QCM	Data	QCM
D^0	$6.819 \pm 0.457^{+0.912}_{-0.936} \pm 0.054$	8.438	$1.275 \pm 0.099^{+0.167}_{-0.173} \pm 0.010$	1.436	—	0.157
D^+	$3.041 \pm 0.073^{+0.154}_{-0.155} \pm 0.052^{+0.352}_{-0.618}$	3.563	$0.552 \pm 0.008^{+0.024}_{-0.024} \pm 0.009^{+0.068}_{-0.114}$	0.606	—	0.0665
D^{*+}	$3.803 \pm 0.037^{+0.084}_{-0.085} \pm 0.041^{+0.854}_{-1.175}$	3.600	$0.663 \pm 0.023^{+0.038}_{-0.039} \pm 0.007^{+0.149}_{-0.165}$	0.613	—	0.0672
D_s^+	$1.89 \pm 0.07^{+0.13+0.36}_{-0.16-0.55} \pm 0.07$	2.417	$0.34 \pm 0.01^{+0.02+0.11}_{-0.03-0.09} \pm 0.01$	0.395	—	0.0368
Λ_c^+	—	5.983	—	1.026	—	0.115
Σ_c^0	—	0.997	—	0.171	—	0.0192
Σ_c^{++}	—	0.997	—	0.171	—	0.0192
Ξ_c^0	—	1.211	—	0.199	—	0.0189
Ξ_c^+	—	1.211	—	0.199	—	0.0189
Ω_c^0	—	0.246	—	0.0386	—	0.00311

4.2. Yield Ratios for Charm Hadrons

In this subsection, we calculate two kinds of yield ratios as functions of p_T for charm hadrons: one is D_s^+/D^0 which is related to strangeness production, and the other is the baryon-to-meson ratio.

We first look at the results for D_s^+/D^0 in Figure 7 at 0–10%, 30–50% and 60–80% centralities. The symbols in panels (a) and (b) are from the most recent data [85], while those in panel (c) are from previous measurements [31]. Different lines are the QCM results. The agreement between the data and our results with the same value of λ_s extracted from strange hadrons implies the same 'strangeness' environment for both strange and charm hadrons and supports the QCM works as the hadronization mechanism for charm quarks in the QGP medium.

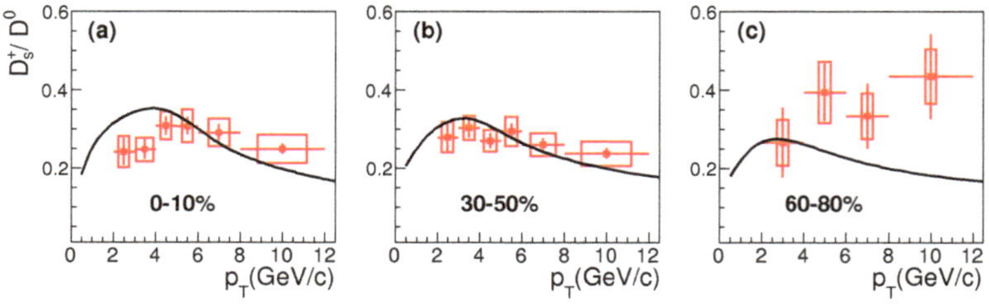

Figure 7. D_s^+/D^0 as a function of p_T for (**a**) 0–10%, (**b**) 30–50%, and (**c**) 60–80% centralities. The symbols are experimental data [31,85], and solid lines are the QCM results.

We then look at the baryon-to-meson ratio Λ_c^+/D^0, which is considered a probe to the charm quark hadronization. Recalling Equations (17,20), we have

$$\frac{\Lambda_c^+}{D^0} = \frac{4.267}{2+\lambda_s} \cdot \frac{\mathcal{A}_B}{\mathcal{A}_M} \cdot \frac{[f_d^{(n)}(x_{ddc}^d p_T)]^2 f_c^{(n)}(x_{ddc}^c p_T)}{f_d^{(n)}(x_{dc}^d p_T) f_c^{(n)}(x_{dc}^c p_T)}. \tag{29}$$

In our calculation, we set $\mathcal{A}_B/\mathcal{A}_M$ to 0.45 by $R_{B/M} = 0.6$ in the charm sector [75]. Following Equation (29), the results for Λ_c^+/D^0 as a function of p_T at different centralities are given in Figure 8. We see that Λ_c^+/D^0 as a function of p_T shows a similar shape to Λ/K_S^0 in Figure 3 except a larger shift in the p_T at the peak value from the central to peripheral collisions; in the central to peripheral collisions, the peak values decrease from about 1.3 to 0.9 and their locations in the p_T shift from 5 to 3 GeV. The peak locations shifting lower p_T from the central to peripheral collisions is due to the stronger collectivity in more central collisions.

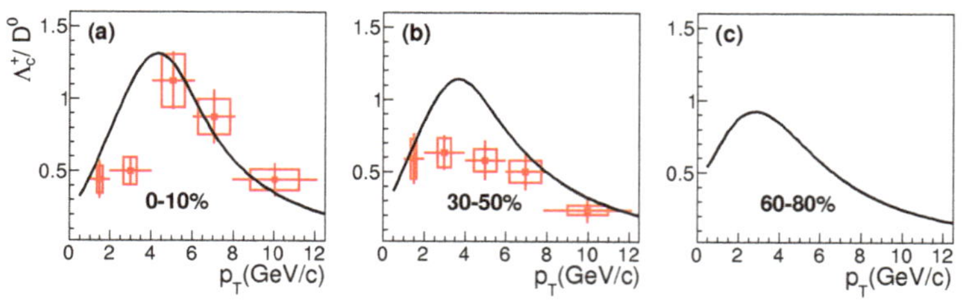

Figure 8. The baryon-to-meson ratio Λ_c^+/D^0 as a function of p_T for (**a**) 0–10%, (**b**) 30–50%, and (**c**) 60–80% centralities. Symbols are experimental data [86], and solid lines are the QCM results.

4.3. Nuclear Modification Factor R_{AA}

We finally investigate the nuclear modification factor R_{AA} for charm hadrons, which is defined as

$$R_{AA}(p_T) = \frac{1}{\langle T_{AA}\rangle}\frac{dN_{AA}/dp_T}{d\sigma_{pp}/dp_T}. \tag{30}$$

The results for differential cross-sections of charm hadrons in pp collisions at $\sqrt{s} = 5.02$ TeV are taken from reference [77] by some of us. Figure 9 shows R_{AA} for prompt D^0, D^+ and D^{*+} mesons as well as Λ_c^+ at 0–10% and 30–50% centralities. Symbols are experimental data [84,86], and solid lines are the QCM results. One can see in Figure 9 that both D mesons and Λ_c^+ have similar p_T behaviors. The peaks are located at p_T^{peak} which shifts towards higher values from peripheral to central collisions. This shift is mainly due to the stronger collectivity in central collisions which can boost thermal quarks to larger transverse momenta that are passed to charm hadrons by the EVC mechanism. We also see that the peak shift for R_{AA} of Λ_c^+ is more obvious than that of D mesons, because Λ_c^+ contains two light quarks and therefore is more influenced by centrality-dependent collectivity.

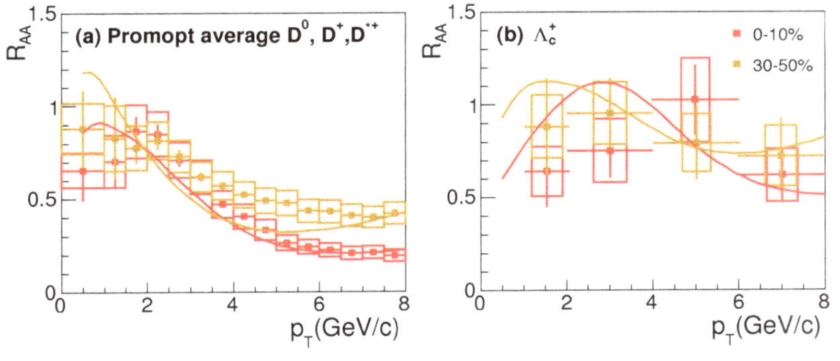

Figure 9. The nuclear modification factor R_{AA} for (**a**) prompt D^0, D^+ and D^{*+} mesons and (**b**) Λ_c^+ baryons as functions of p_T at different centralities. Symbols are experimental data from Refs. [84,86], and solid lines are the QCM results.

5. Summary

The comparative study of the production properties of strange and charm hadrons can provide information on the hadronization mechanism in relativistic heavy ion collisions. We use a quark combination model in momentum space with the approximation of equal-velocity combination to study these properties in Pb+Pb collisions at $\sqrt{s_{NN}} = 5.02$ TeV. We used experimental data of Λ, ϕ and D^0 to fix the p_T spectra of up, strange, and charm quarks at hadronization. We computed the p_T spectra and rapidity densities of K_s^0, Ξ^-, Ω^-, D^+, D_s^+, D^{*+}, Λ_c^+, Σ_c^0, Ξ_c^+ and Ω_c^0 from central to peripheral collisions. The QCM results agree with the available experimental data quite well.

We calculated the yield ratios Λ/K_S^0, D_s^+/D^0 and Λ_c^+/D^0 as functions of p_T. We found that the EVC-based QCM in momentum space can naturally describe their non-trivial behaviors as functions of the p_T and the centrality. The p_T locations of the peaks in these ratio curves shift to lower values from central to peripheral collisions, an effect arising from collectivity in heavy ion collisions, absent in pp collisions. The calculated results for the nuclear modification factor R_{AA} show similar behaviors for both D and Λ_c^+, which can be tested by more precise experimental measurements, especially at a low p_T. All our results support the validity of the EVC-based QCM in describing the hadronization mechanism of charm quarks in high energy heavy ion collisions.

Author Contributions: Conceptualization, F.-L.S. and Z.-T.L.; methodology, J.S., F.-L.S. and Z.-T.L.; formal analysis, R.-Q.W. and Q.W.; investigation, J.S., W.-B.C. and R.-Q.W.; data curation, W.-B.C.; writing—original draft preparation, R.-Q.W.; writing—review and editing, Q.W. and F.-L.S.; supervision, F.-L.S. All authors have read and agreed to the published version of the manuscript.

Funding: This research was funded by the National Natural Science Foundation of China under grant nos. 12175115, 11975011, 12135011, 11890710 and 11890713, and the Natural Science Foundation of Shandong Province, China, under grant nos. ZR2019MA053 and ZR2020MA097.

Institutional Review Board Statement: Not applicable.

Informed Consent Statement: Not applicable.

Data Availability Statement: Not applicable.

Acknowledgments: We dedicate this work to Qu-Bing Xie (1935–2013) who was the teacher, mentor, and friend of Z.-T.L., F.-L.S. and Q.W.

Conflicts of Interest: The authors declare no conflict of interest.

References

1. Shuryak, E.V. Quantum Chromodynamics and the Theory of Superdense Matter. *Phys. Rept.* **1980**, *61*, 71–158. [CrossRef]
2. Braun-Munzinger, P.; Wambach, J. The Phase Diagram of Strongly-Interacting Matter. *Rev. Mod. Phys.* **2009**, *81*, 1031–1050. [CrossRef]
3. Fukushima, K.; Hatsuda, T. The phase diagram of dense QCD. *Rept. Prog. Phys.* **2011**, *74*, 014001. [CrossRef]
4. Borsanyi, S.; Fodor, Z.; Hoelbling, C.; Katz, S.D.; Krieg, S.; Ratti, C.; Szabo, K.K. Is there still any T_c mystery in lattice QCD? Results with physical masses in the continuum limit III. *JHEP* **2010**, *09*, 073. [CrossRef]
5. Bazavov, A.; Bhattacharya, T.; Cheng, M.; DeTar, C.; Ding, H.-T.; Gottlieb, S.; Gupta, R.; Hegde, P.; Heller, U.M.; Karsch, F.; et al. The chiral and deconfinement aspects of the QCD transition. *Phys. Rev. D* **2012**, *85*, 054503. [CrossRef]
6. Ding, H.T.; Karsch, F.; Mukherjee, S. Thermodynamics of strong-interaction matter from Lattice QCD. *Int. J. Mod. Phys. E* **2015**, *24*, 1530007. [CrossRef]
7. Adams, J.; Aggarwal, M.M.; Ahammed, Z.; Amonett, J.; Anderson, B.D.; Arkhipkin, D.; Averichev, G.S.; Badyal, S.K.; Bai, Y.; Balewski, J.; et al. Experimental and theoretical challenges in the search for the quark gluon plasma: The STAR Collaboration's critical assessment of the evidence from RHIC collisions. *Nucl. Phys. A* **2005**, *757*, 102–183. [CrossRef]
8. Adcox, K.; Adler, S.S.; Afanasiev, S.; Aidala, C.; Ajitanand, N.N.; Akiba, Y.; Al-Jamel, A.; Alexander, J.; Amirikas, R.; Aoki, K.; et al. Formation of dense partonic matter in relativistic nucleus-nucleus collisions at RHIC: Experimental evaluation by the PHENIX collaboration. *Nucl. Phys. A* **2005**, *757*, 184–283. [CrossRef]
9. Back, B.B.; Baker, M.D.; Ballintijn, M.; Barton, D.S.; Becker, B.; Betts, R.R.; Bickley, A.A.; Bindel, R.; Budzanowski, A.; Busza, W.; et al. The PHOBOS perspective on discoveries at RHIC. *Nucl. Phys. A* **2005**, *757*, 28–101. [CrossRef]
10. Arsene, I.; Bearden, I.G.; Beavis, D.; Besliu, C.; Budick, B.; Boggild, H.; Chasman, C.; Christensen, C.H.; Christiansen, P.; Cibor, J.; et al. Quark gluon plasma and color glass condensate at RHIC? The Perspective from the BRAHMS experiment. *Nucl. Phys. A* **2005**, *757*, 1–27. [CrossRef]
11. Rafelski, J.; Muller, B. Strangeness Production in the Quark-Gluon Plasma. *Phys. Rev. Lett.* **1986**, *48*, 1066. Erratum in *Phys. Rev. Lett.* **1986**, *56*, 2334. [CrossRef]
12. Shor, A. Phi meson production as a probe of the quark gluon plasma. *Phys. Rev. Lett.* **1985**, *54*, 1122–1125. [CrossRef] [PubMed]
13. van Hecke, H.; Sorge, H.; Xu, N. Evidence of early multistrange hadron freezeout in high-energy nuclear collisions. *Phys. Rev. Lett.* **1998**, *81*, 5764–5767. [CrossRef]
14. Matsui, T.; Satz, H. J/ψ Suppression by Quark-Gluon Plasma Formation. *Phys. Lett. B* **1986**, *178*, 416–422. [CrossRef]
15. van Hees, H.; Greco, V.; Rapp, R. Heavy-quark probes of the quark-gluon plasma at RHIC. *Phys. Rev. C* **2006**, *73*, 034913. [CrossRef]
16. Mocsy, A.; Petreczky, P. Color screening melts quarkonium. *Phys. Rev. Lett.* **2007**, *99*, 211602. [CrossRef]
17. He, M.; Fries, R.J.; Rapp, R. D_s-Meson as Quantitative Probe of Diffusion and Hadronization in Nuclear Collisions. *Phys. Rev. Lett.* **2013**, *110*, 112301. [CrossRef]
18. Braun-Munzinger, P.; Koch, V.; Schäfer, T.; Stachel, J. Properties of hot and dense matter from relativistic heavy ion collisions. *Phys. Rept.* **2016**, *621*, 76–126. [CrossRef]
19. Batsouli, S.; Kelly, S.; Gyulassy, M.; Nagle, J.L. Does the charm flow at RHIC? *Phys. Lett. B* **2003**, *557*, 26–32. [CrossRef]
20. Xu, N.; Xu, Z. Transverse dynamics at RHIC. *Nucl. Phys. A* **2003**, *715*, 587–590. [CrossRef]
21. Lin, Z.W.; Molnar, D. Quark coalescence and elliptic flow of charm hadrons. *Phys. Rev. C* **2003**, *68*, 044901. [CrossRef]
22. Moore, G.D.; Teaney, D. How much do heavy quarks thermalize in a heavy ion collision? *Phys. Rev. C* **2005**, *71*, 064904. [CrossRef]
23. Abelev, B.I.; Aggarwal, M.M.; Ahammed, Z.; Anderson, B.D.; Arkhipkin, D.; Averichev, G.S.; Bai, Y.; Balewski, J.; Barannikova, O.; Barnby, L.S.; et al. Measurements of phi meson production in relativistic heavy-ion collisions at RHIC. *Phys. Rev. C* **2009**, *79*, 064903. [CrossRef]

24. Adamczyk, L.; Adkins, J.K.; Agakishiev, G.; Aggarwal, M.M.; Ahammed, Z.; Alekseev, I.; Alford, J.; Aparin, A.; Arkhipkin, D.; Aschenauer, E.C.; et al. Probing parton dynamics of QCD matter with Ω and ϕ production. *Phys. Rev. C* **2016**, *93*, 021903. [CrossRef]
25. Abelev, B.B.; Adam, J.; Adamova, D.; Adare, A.M.; Aggarwal, M.M.; Rinella, G.A.; Agnello, M.; Agocs, A.G.; Agostinelli, A.; Ahammed, Z.; et al. Multi-strange baryon production at mid-rapidity in Pb-Pb collisions at $\sqrt{s_{NN}}$ = 2.76 TeV. *Phys. Lett. B* **2014**, *728*, 216–227. Erratum in Phys. Lett. B **2014**, *734*, 409–410. [CrossRef]
26. Adler, S.S.; Afanasiev, S.; Aidala, C.; Ajitanand, N.N.; Akiba, Y.; Alexander, J.; Amirikas, R.; Aphecetche, L.; Aronson, S.H.; Averbeck, R.; et al. Centrality dependence of charm production from single electrons measurement in Au + Au collisions at s(NN)**(1/2) = 200-GeV. *Phys. Rev. Lett.* **2005**, *94*, 082301. [CrossRef] [PubMed]
27. Adams, J.; Aggarwal, M.M.; Ahammed, Z.; Amonett, J.; Anderson, B.D.; Arkhipkin, D.; Averichev, G.S.; Badyal, S.K.; Bai, Y.; Balewski, J.; et al. Open charm yields in d + Au collisions at s(NN)**(1/2) = 200-GeV. *Phys. Rev. Lett.* **2005**, *94*, 062301. [CrossRef]
28. Abelev, B.; Adam, J.; Adamova, D.; Adare, A.M.; Aggarwal, M.; Rinella, G.A.; Agocs, A.G.; Agostinelli, A.; Salazar, S.A.; Ahammed, Z.; et al. Measurement of charm production at central rapidity in proton-proton collisions at $\sqrt{s} = 2.76$ TeV. *JHEP* **2012**, *7*, 191. [CrossRef]
29. Abelev, B.; Adam, J.; Adamova, D.; Adare, A.M.; Aggarwal, M.; Rinella, G.A.; Agocs, A.G.; Agostinelli, A.; Salazar, S.A.; Ahammed, Z.; et al. Suppression of high transverse momentum D mesons in central Pb-Pb collisions at $\sqrt{s_{NN}} = 2.76$ TeV. *JHEP* **2012**, *9*, 112. [CrossRef]
30. Adam, J.; Adamczyk, L.; Adams, J.; Adkins, K.; Agakishiev, G.; Aggarwal, M.; Ahammed, Z.; Alekseev, I.; Anderson, D.; Aoyama, R.; et al. Centrality and transverse momentum dependence of D^0-meson production at mid-rapidity in Au+Au collisions at $\sqrt{s_{NN}} = 200$ GeV. *Phys. Rev. C* **2019**, *99*, 034908. [CrossRef]
31. Acharya, S.; Acosta, F.T.; Adamová, D.; Adolfsson, J.; Aggarwal, M.M.; Rinella, G.A.; Agnello, M.; Agrawal, N.; Ahammed, Z.; Ahn, S.U.; et al. Measurement of D^0, D^+, D^{*+} and D_s^+ production in Pb-Pb collisions at $\sqrt{s_{NN}} = 5.02$ TeV. *JHEP* **2018**, *10*, 174. [CrossRef]
32. Radhakrishnan, S. Measurements of open charm production in Au+Au collisions at $\sqrt{s_{NN}}$ = 200GeV with the STAR experiment at RHIC. *Nucl. Phys. A* **2019**, *982*, 659–662. [CrossRef]
33. Zhou, L. Measurements of Λ_c^+ and D_s^+ productions in Au+Au collisions at $\sqrt{s_{NN}}$ = 200 GeV from STAR. *Nucl. Phys. A* **2017**, *967*, 620–623. [CrossRef]
34. Acharya, S.; Adamová, D.; Adhya, S.P.; Adler, A.; Adolfsson, J.; Aggarwal, M.M.; Rinella, G.A.; Agnello, M.; Agrawal, N.; Ahammed, Z.; et al. Measurement of prompt D^0, D^+, D^{*+}, and D_S^+ production in p–Pb collisions at $\sqrt{s_{NN}}$ = 5.02 TeV. *JHEP* **2019**, *12*, 092. [CrossRef]
35. Acharya, S.; Acosta, F.T.; Adamova, D.; Adhya, S.P.; Adler, A.; Adolfsson, J.; Aggarwal, M.M.; Rinella, G.A.; Agnello, M.; Agrawal, N.; et al. Λ_c^+ production in Pb-Pb collisions at $\sqrt{s_{NN}} = 5.02$ TeV. *Phys. Lett. B* **2019**, *793*, 212–223. [CrossRef]
36. Acharya, S.; Acosta, F.T.; Adamová, D.; Adolfsson, J.; Aggarwal, M.M.; Rinella, G.A.; Agnello, M.; Agrawal, N.; Ahammed, Z.; Ahn, S.U.; et al. Λ_c^+ production in pp collisions at $\sqrt{s} = 7$ TeV and in p-Pb collisions at $\sqrt{s_{NN}} = 5.02$ TeV. *JHEP* **2018**, *04*, 108. [CrossRef]
37. Adams, J.; Aggarwal, M.M.; Ahammed, Z.; Amonett, J.; Anderson, B.D.; Anderson, M.; Arkhipkin, D.; Averichev, G.S.; Bai, Y.; Balewski, J.; et al. Scaling Properties of Hyperon Production in Au+Au Collisions at s**(1/2) = 200-GeV. *Phys. Rev. Lett.* **2007**, *98*, 062301. [CrossRef]
38. Adams, J.; Aggarwal, M.M.; Ahammed, Z.; Amonett, J.; Anderson, B.D.; Anderson, M.; Arkhipkin, D.; Averichev, G.S.; Bai, Y.; Balewski, J.; et al. Bulk Properties of the Medium Produced in Relativistic Heavy-Ion Collisions from the Beam Energy Scan Program. *Phys. Rev. C* **2017**, *96*, 044904. [CrossRef]
39. Adam, J.; Adamczyk, L.; Adams, J.; Adkins, K.; Agakishiev, G.; Aggarwal, M.; Ahammed, Z.; Alekseev, I.; Anderson, D.; Aoyama, R.; et al. Strange hadron production in Au+Au collisions at $\sqrt{s_{NN}}$ = 7.7, 11.5, 19.6, 27, and 39 GeV. *Phys. Rev. C* **2020**, *102*, 034909. [CrossRef]
40. Acharya, S.; Adamova, D.; Adler, A.; Adolfsson, J.; Rinella, G.A.; Agnello, M.; Agrawal, N.; Ahammed, Z.; Ahmad, S.; Ahn, S.U.; et al. Production of $K^*(892)^0$ and $\phi(1020)$ in pp and Pb-Pb collisions at $\sqrt{s_{NN}} = 5.02$ TeV. *Phys. Rev. C* **2022**, *106*, 034907. [CrossRef]
41. Abelev, B.I.; Aggarwal, M.M.; Ahammed, Z.; Anderson, B.D.; Anderson, M.; Arkhipkin, D.; Averichev, G.S.; Bai, Y.; Balewski, J.; Barannikova, O.; et al. Identified baryon and meson distributions at large transverse momenta from Au+Au collisions at s(NN)**(1/2) = 200-GeV. *Phys. Rev. Lett.* **2006**, *97*, 152301. [CrossRef]
42. Adare, A.; Afanasiev, S.; Aidala, C.; Ajitanand, N.N.; Akiba, Y.; Al-Bataineh, H.; Alexander, J.; Al-Jamel, A.; Aoki, K.; Aphecetche, L.; et al. Scaling properties of azimuthal anisotropy in Au+Au and Cu+Cu collisions at s(NN) = 200-GeV. *Phys. Rev. Lett.* **2007**, *98*, 162301. [CrossRef] [PubMed]
43. Blume, C.; Markert, C. Strange hadron production in heavy ion collisions from SPS to RHIC. *Prog. Part. Nucl. Phys.* **2011**, *66*, 834–879. [CrossRef]
44. Abelev, B.B.; Adam, J.; Adamova, D.; Aggarwal, M.M.; Rinella, G.A.; Agnello, M.; Agostinelli, A.; Agrawal, N.; Ahammed, Z.; Ahmad, N.; et al. Production of charged pions, kaons and protons at large transverse momenta in pp and Pb–Pb collisions at $\sqrt{s_{NN}}$ =2.76 TeV. *Phys. Lett. B* **2014**, *736*, 196–207. [CrossRef]

45. Kolb, P.F.; Heinz, U.W.; Huovinen, P.; Eskola, K.J.; Tuominen, K. Centrality dependence of multiplicity, transverse energy, and elliptic flow from hydrodynamics. *Nucl. Phys. A* **2001**, *696*, 197–215. [CrossRef]
46. Shen, C.; Heinz, U.; Huovinen, P.; Song, H. Systematic parameter study of hadron spectra and elliptic flow from viscous hydrodynamic simulations of Au+Au collisions at $\sqrt{s_{NN}} = 200$ GeV. *Phys. Rev. C* **2010**, *82*, 054904. [CrossRef]
47. Andronic, A.; Braun-Munzinger, P.; Redlich, K.; Stachel, J. Charmonium and open charm production in nuclear collisions at SPS/FAIR energies and the possible influence of a hot hadronic medium. *Phys. Lett. B* **2008**, *659*, 149–155. [CrossRef]
48. Kuznetsova, I.; Rafelski, J. heavy-flavor hadrons in statistical hadronization of strangeness-rich QGP. *Eur. Phys. J. C* **2007**, *51*, 113–133. [CrossRef]
49. Andronic, A.; Arleo, F.; Arnaldi, R.; Beraudo, A.; Bruna, E.; Caffarri, D.; Valle, Z.C.D.; Contreras, J.G.; Dahms, T.; Dainese, A.; et al. Heavy-flavour and quarkonium production in the LHC era: From proton–proton to heavy-ion collisions. *Eur. Phys. J. C* **2016**, *76*, 107. [CrossRef]
50. Greco, V.; Ko, C.M.; Lévai, P. Parton coalescence and anti-proton/pion anomaly at RHIC. *Phys. Rev. Lett.* **2003**, *90*, 202302. [CrossRef]
51. Fries, R.J.; Müller, B.; Nonaka, C.; Bass, S.A. Hadronization in heavy ion collisions: Recombination and fragmentation of partons. *Phys. Rev. Lett.* **2003**, *90*, 202303. [CrossRef] [PubMed]
52. Hwa, R.C.; Yang, C.B. Scaling behavior at high p(T) and the p/pi ratio. *Phys. Rev.* **2003**, *C67*, 034902. [CrossRef]
53. Shao, F.l.; Xie, Q.b.; Wang, Q. Production rates for hadrons, pentaquarks Theta+ and Theta*++, and di-baryon (Omega Omega)(0+) in relativistic heavy ion collisions by a quark combination model. *Phys. Rev. C* **2005**, *71*, 044903. [CrossRef]
54. Greco, V.; Ko, C.M.; Rapp, R. Quark coalescence for charmed mesons in ultrarelativistic heavy ion collisions. *Phys. Lett. B* **2004**, *595*, 202–208. [CrossRef]
55. Oh, Y.; Ko, C.M.; Lee, S.H.; Yasui, S. Heavy baryon/meson ratios in relativistic heavy ion collisions. *Phys. Rev. C* **2009**, *79*, 044905. [CrossRef]
56. Liu, Y.; Greiner, C.; Kostyuk, A. B_c meson enhancement and the momentum dependence in Pb + Pb collisions at energies available at the CERN Large Hadron Collider. *Phys. Rev. C* **2013**, *87*, 014910. [CrossRef]
57. Lee, S.H.; Ohnishi, K.; Yasui, S.; Yoo, I.K.; Ko, C.M. Lambda(c) enhancement from strongly coupled quark-gluon plasma. *Phys. Rev. Lett.* **2008**, *100*, 222301. [CrossRef]
58. He, M.; Fries, R.J.; Rapp, R. Heavy-Quark Diffusion and Hadronization in Quark-Gluon Plasma. *Phys. Rev. C* **2012**, *86*, 014903. [CrossRef]
59. Plumari, S.; Minissale, V.; Das, S.K.; Coci, G.; Greco, V. Charmed Hadrons from Coalescence plus Fragmentation in relativistic nucleus-nucleus collisions at RHIC and LHC. *Eur. Phys. J. C* **2018**, *78*, 348. [CrossRef]
60. Zhao, J.; Shi, S.; Xu, N.; Zhuang, P. Sequential Coalescence with Charm Conservation in High Energy Nuclear Collisions. *arXiv* **2018**, arXiv:1805.10858.
61. Cho, S.; Sun, K.J.; Ko, C.M.; Lee, S.H.; Oh, Y. Charmed hadron production in an improved quark coalescence model. *Phys. Rev. C* **2020**, *101*, 024909. [CrossRef]
62. Wang, R.q.; Song, J.; Shao, F.l. Yield correlations and p_T dependence of charm hadrons in Pb + Pb collisions at $\sqrt{s_{NN}} = 2.76$ TeV. *Phys. Rev. C* **2015**, *91*, 014909. [CrossRef]
63. He, M.; Fries, R.J.; Rapp, R. heavy-flavor at the Large Hadron Collider in a Strong Coupling Approach. *Phys. Lett. B* **2014**, *735*, 445–450. [CrossRef]
64. Molnar, D.; Voloshin, S.A. Elliptic flow at large transverse momenta from quark coalescence. *Phys. Rev. Lett.* **2003**, *91*, 092301. [CrossRef]
65. Xie, Q.B.; Mo, W.C.; Li, Y.F. THE COMBINATION RULE OF QUARKS IN E+ E- ANNIHILATION. (IN CHINESE). *HEPNP* **1984**, *8*, 642–647.
66. Xie, Q.B.; Liu, X.M. Quark Production Rule in $e^+e^- \to$ Two Jets. *Phys. Rev. D* **1988**, *38*, 2169–2177. [CrossRef]
67. Xie, Q.B. Quark combination rule and correlations between baryons. In Proceedings of the 19th International Symposium on Multiparticle Dynamics: New Data and Theoretical Trends, Arles, France, 13–17 June 1988.
68. Chen, C.J.; Ma, W.J.; Xie, Q.B. THE NEAREST CORRELATION IN RAPIDITY AND THE MULTIPLICITY DISTRIBUTION IN E+ E- ANNIHILATION. *J. Phys. G* **1988**, *14*, 1339–1344. [CrossRef]
69. Fang, H.P.; Xie, Q.B.; Lai, X.P. RAPIDITY DISTRIBUTION AND BARYON-ANTI-BARYON CORRELATIONS IN E+ E- —> TWO JET EVENTS. (IN CHINESE). *HEPNP* **1989**, *13*, 518–525.
70. Liang, Z.T.; Xie, Q.B. Baryon anti-baryon flavor correlations in e+ e- annihilation. *Phys. Rev. D* **1991**, *43*, 751–759. [CrossRef]
71. Shao, C.E.; Song, J.; Shao, F.L.; Xie, Q.B. Hadron production by quark combination in central Pb+Pb collisions at s(NN)**(1/2) = 17.3-GeV. *Phys. Rev.* **2009**, *C80*, 014909. [CrossRef]
72. Song, J.; Shao, F.L. Baryon-antibaryon production asymmetry in relativistic heavy ion collisions. *Phys. Rev. C* **2013**, *88*, 027901. [CrossRef]
73. Song, J.; Gou, X.R.; Shao, F.L.; Liang, Z.T. Quark number scaling of hadronic p_T spectra and constituent quark degree of freedom in *p*-Pb collisions at $\sqrt{s_{NN}} = 5.02$ TeV. *Phys. Lett.* **2017**, *B774*, 516–521. [CrossRef]
74. Gou, X.R.; Shao, F.L.; Wang, R.Q.; Li, H.H.; Song, J. New insights into hadron production mechanism from p_T spectra in *pp* collisions at $\sqrt{s} = 7$ TeV. *Phys. Rev.* **2017**, *D96*, 094010. [CrossRef]

75. Wang, R.Q.; Song, J.; Shao, F.L.; Liang, Z.T. Charmed hadron production via equal-velocity quark combination in ultrarelativistic heavy ion collisions. *Phys. Rev. C* **2020**, *101*, 054903. [CrossRef]
76. Song, J.; Shao, F.L.; Liang, Z.T. Quark number scaling of p_T spectra for Ω and ϕ in relativistic heavy-ion collisions. *Phys. Rev. C* **2020**, *102*, 014911. [CrossRef]
77. Li, H.H.; Shao, F.L.; Song, J. Production of light-flavor and single-charmed hadrons in pp collisions at $\sqrt{s}$ = 5.02 TeV in an equal-velocity quark combination model. *Chin. Phys. C* **2021**, *45*, 113105. [CrossRef]
78. Song, J.; Li, H.H.; Shao, F.L. Signals of quark combination at hadronization in pp collisions at $\sqrt{s}$ = 200 GeV. *Phys. Rev. D* **2022**, *105*, 074027. [CrossRef]
79. Song, J.; Li, H.H.; Shao, F.L. New feature of low p_T charm quark hadronization in pp collisions at $\sqrt{s} = 7$ TeV. *Eur. Phys. J. C* **2018**, *78*, 344. [CrossRef]
80. Li, H.H.; Shao, F.L.; Song, J.; Wang, R.Q. Production of single-charm hadrons by quark combination mechanism in p-Pb collisions at $\sqrt{s_{NN}} = 5.02$ TeV. *Phys. Rev.* **2018**, *C97*, 064915. [CrossRef]
81. Acharya, S.; Adamova, D.; Adler, A.; Adolfsson, J.; Rinella, G.A.; Agnello, M.; Agrawal, N.; Ahammed, Z.; Ahmad, S.; Ahn, S.U.; et al. Measurement of Prompt D^0, Λ_c^+, and $\Sigma_c^{0,++}$(2455) Production in Proton–Proton Collisions at $\sqrt{s}$ = 13 TeV. *Phys. Rev. Lett.* **2022**, *128*, 012001. [CrossRef]
82. First measurement of Λ_c^+ production down to $p_T = 0$ in pp and p-Pb collisions at $\sqrt{s_{NN}} = 5.02$ TeV. *arXiv* **2022**, arXiv:2211.14032.
83. Acharya, S.; Adamova, D.; Adler, A.; Adolfsson, J.; Rinella, G.A.; Agnello, M.; Agrawal, N.; Ahammed, Z.; Ahmad, S.; Ahn, S.U.; et al. Inclusive heavy-flavour production at central and forward rapidity in Xe-Xe collisions at $\sqrt{s_{NN}}$ = 5.44 TeV. *Phys. Lett. B* **2021**, *819*, 136437. [CrossRef]
84. Acharya, S.; Adamová, D.; Adler, A.; Adolfsson, J.; Rinella, G.A.; Agnello, M.; Agrawal, N.; Ahammed, Z.; Ahmad, S.; Ahn, S.U.; et al. Prompt D^0, D^+, and D^{*+} production in Pb–Pb collisions at $\sqrt{s_{NN}}$ = 5.02 TeV. *JHEP* **2022**, *1*, 174. [CrossRef]
85. Acharya, S.; Adamova, D.; Adler, A.; Adolfsson, J.; Rinella, G.A.; Agnello, M.; Agrawal, N.; Ahammed, Z.; Ahmad, S.; Ahn, S.U.; et al. Measurement of prompt D_s^+-meson production and azimuthal anisotropy in Pb–Pb collisions at $\sqrt{s_{NN}}$ = 5.02TeV. *Phys. Lett. B* **2022**, *827*, 136986. [CrossRef]
86. Acharya, S.; Adamova, D.; Adler, A.; Adolfsson, J.; Rinella, G.A.; Agnello, M.; Agrawal, N.; Ahammed, Z.; Ahmad, S.; Ahn, S.U.; et al. Constraining hadronization mechanisms with Λ_c^+/D^0 production ratios in Pb-Pb collisions at $\sqrt{s_{NN}} = 5.02$ TeV. *arXiv* **2021**, arXiv:2112.08156.
87. Yang, Y.g.; Song, J.; Shao, F.l.; Liang, Z.t.; Wang, Q. Statistical method in quark combination model. *Chin. Phys. C* **2020**, *44*, 034103. [CrossRef]
88. Wang, Q.; Xie, Q.B. Production percentages of excited baryons in e+ e- annihilation. *J. Phys. G* **1995**, *21*, 897–904. [CrossRef]
89. Wang, R.q.; Shao, F.l.; Song, J.; Xie, Q.b.; Liang, Z.t. Hadron Yield Correlation in Combination Models in High Energy AA Collisions. *Phys. Rev. C* **2012**, *86*, 054906. [CrossRef]
90. Liang, Z.T.; Wang, X.N. Spin alignment of vector mesons in non-central A+A collisions. *Phys. Lett. B* **2005**, *629*, 20–26. [CrossRef]
91. Yang, Y.G.; Fang, R.H.; Wang, Q.; Wang, X.N. Quark coalescence model for polarized vector mesons and baryons. *Phys. Rev. C* **2018**, *97*, 034917. [CrossRef]
92. Sheng, X.L.; Oliva, L.; Wang, Q. What can we learn from the global spin alignment of ϕ mesons in heavy-ion collisions? *Phys. Rev. D* **2022**, *101*, 096005. [CrossRef]
93. Sheng, X.L.; Oliva, L.; Liang, Z.T.; Wang, Q.; Wang, X.N. Spin alignment of vector mesons in heavy-ion collisions. *arXiv* **2022**, arXiv:2205.15689.
94. Wiedemann, U.A.; Scotto, P.; Heinz, U.W. Transverse momentum dependence of Hanbury-Brown-Twiss correlation radii. *Phys. Rev. C* **1996**, *53*, 918–931. [CrossRef] [PubMed]
95. Retiere, F.; Lisa, M.A. Observable implications of geometrical and dynamical aspects of freeze out in heavy ion collisions. *Phys. Rev. C* **2004**, *70*, 044907. [CrossRef]
96. Tanabashi, M.; Hagiwara, K.; Hikasa, K.; Nakamura, K.; Sumino, Y.; Takahashi, F.; Tanaka, J.; Agashe, K.; Aielli, G.; Amsler, C.; et al. Review of Particle Physics. *Phys. Rev. D* **2018**, *98*, 030001. [CrossRef]
97. Acharya, S.; Adamova, D.; Adler, A.; Adolfsson, J.; Aggarwal, M.M.; Rinella, G.A.; Agnello, M.; Agrawal, N.; Ahammed, Z.; Ahmad, S.; et al. Evidence of rescattering effect in Pb-Pb collisions at the LHC through production of $K^*(892)^0$ and $\phi(1020)$ mesons. *Phys. Lett. B* **2020**, *802*, 135225. [CrossRef]
98. Kalinak, P. Strangeness production in Pb-Pb collisions with ALICE at the LHC. *PoS* **2017**, *EPS-HEP2017*, 168. [CrossRef]
99. Greco, V.; Ko, C.M.; Lévai, P. Parton coalescence at RHIC. *Phys. Rev. C* **2003**, *68*, 034904. [CrossRef]
100. Fries, R.J.; Muller, B.; Nonaka, C.; Bass, S.A. Hadron production in heavy ion collisions: Fragmentation and recombination from a dense parton phase. *Phys. Rev. C* **2003**, *68*, 044902. [CrossRef]
101. Scardina, F.; Das, S.K.; Minissale, V.; Plumari, S.; Greco, V. Estimating the charm quark diffusion coefficient and thermalization time from D meson spectra at energies available at the BNL Relativistic Heavy Ion Collider and the CERN Large Hadron Collider. *Phys. Rev. C* **2017**, *96*, 044905. [CrossRef]

symmetry

MDPI

Article

On Thermodynamically Consistent Quasiparticle Model at Finite Chemical Potential †

Wei-Liang Qian [1,2,3,*], Hong-Hao Ma [4,5,*], Shaoyu Yin [6] and Ping Wang [7]

1 Escola de Engenharia de Lorena, Universidade de São Paulo, Lorena 12602-810, SP, Brazil
2 Faculdade de Engenharia de Guaratinguetá, Universidade Estadual Paulista, Guaratinguetá 12516-410, SP, Brazil
3 Center for Gravitation and Cosmology, College of Physical Science and Technology, Yangzhou University, Yangzhou 225009, China
4 Department of Physics, Guangxi Normal University, Guilin 541004, China
5 Guangxi Key Laboratory of Nuclear Physics and Technology, Guangxi Normal University, Guilin 541004, China
6 Institute for Theoretical Physics & Cosmology, Zhejiang University of Technology, Hangzhou 310032, China
7 Institute of High Energy Physics, CAS, P.O. Box 918(4), Beijing 100049, China
* Correspondence: wlqian@usp.br (W.-L.Q.); mahonghao@pku.edu.cn (H.-H.M.)
† This article is dedicated to the memory of the late Professor Ru-Keng Su.

Abstract: We explore the quasiparticle model at the finite chemical potential related to Ru-Keng Su's distinguished contributions to the topic. In addition, we discuss the recent developments in the model, and in particular, one argues that the effective mass of the quasiparticle might attain a specific form as a function of momentum, in addition to its dependence on the temperature and chemical potential. Unlike the approaches based on the properties of underlying symmetry or the renormalization group, the momentum dependence emerges as a special solution to an integro-differential equation resulting from the underlying thermodynamic consistency. Moreover, this special solution to the problem is shown to be more general than previously explored in the literature. Instead of fitting to the lattice QCD data at the vanishing chemical potential, in this work, we adopt a "bottom-up" approach by assuming some analytic ansatzes that are manifestly thermodynamically consistent. The remaining physical quantities are subsequently derived, and possible implications are also addressed.

Keywords: hydrodynamics; equation of state; quasiparticle

1. Introduction

The transition between the quark-gluon plasma (QGP) and hadronic phases constitutes one of the most prominent problems in high-energy nuclear physics. In the vicinity of such a region, the underlying dynamics are essentially nonperturbative, through which the system undergoes a dramatic change in the number of degrees of freedom. Moreover, even in the QGP phase, the system's thermodynamic properties deviate significantly from those of a non-interacting ideal gas of quarks and gluons. For instance, the lattice quantum chromodynamics (QCD) calculations showed [1] that the system's pressure and energy density undershoot the Stefan–Boltzmann limit by about 15–20%, even at temperatures $T \gtrsim 3T_c$. Moreover, the speed of sound extracted from a lattice QCD is found to be smaller than that of a massless ideal gas. In particular, as the system approaches the transition region, it is observed that the speed of sound varies non-monotonically [2]. The above properties are crucial for adequately establishing the equation of state (EoS), which plays a central role in providing an appropriate description of the hydrodynamic evolution of the hot and dense system that emerged in the relativistic heavy ion collisions [3–15].

In the literature, the density-dependent quark mass was suggested by Fowler, Raha, and Weiner [16] to address the transition between nuclear and quark matter, and the thermal partonic quasiparticle was initially proposed by Peshier, Kampfer, Pavlenko, and

Soff [17] to accommodate the numerical experiments from the lattice QCD. Primarily inspired by its counterparts in other fields of physics, the notion of the quasiparticle is a phenomenological approach aimed at capturing the bulk thermodynamic properties of QGP. The model can be viewed as an effective and simplified imitation of the essence of many existing theoretical efforts, namely the lattice QCD [1], dimensional reduction [18], hard thermal-loop [19], Polyakov-loop model [20], as well as other hadronic-degree-based approaches [21]. It was also speculated that the success of the notion of a quasiparticle degree of freedom will further give rise to novel effective theories from a more fundamental perspective while properly incorporating the nonperturbative aspects of the QCD.

The quasiparticle model interprets the system as a composition of non-interacting quanta which carry the same quantum numbers of quarks and gluons. The medium-dependent quasiparticle mass implements the strong interactions between the elementary degrees of freedom. For the description of gluon plasma, the quasiparticle mass was initially assumed to be merely temperature dependent [17]. As the concept flourished in the literature, a crucial feature of the model was elaborated in a seminal paper by Gorenstein and Yang [22] with respect to its thermodynamic consistency. The authors solved the issue elegantly via the requirement of an exact cancelation between the additional contributions from the temperature-dependent quasiparticle mass and those from the bag constant.

Subsequently, various relevant aspects of the topic were discussed and further developed by, among others [23–27], Ru-Keng Su in collaboration with his students and collaborators [28–44]. In [28], the role of an additional contribution to the thermopotential and its consequential effect on the strange quark matter were explored. A series of studies regarding the quark mass density- and temperature-dependent (QMDTD) model were performed in [29–32]. The temperature dependence of the stable radius of a strangelet was discussed in [29]. The temperature dependence of the bag constant B was explored and shown to cure the divergence that occurred at the vanishing baryon density in the phase diagram for the bulk strange quark matter of the original QMDTD model [30]. A systematic analysis regarding the stability of a strangelet was performed in [31] in the framework of the QMDTD model. It was observed that stable strangelets are more likely to be encountered in the region with a sizeable negative electric charge and significant strangeness. The analysis was then extended to the dibaryon systems [32] regarding different decay channels, and the results were found in good agreement with those obtained by the chiral SU(3) quark model. The QMDTD setup was then applied to the context of the Friedberg–Lee soliton bag [45–47] nonlinear coupled to the sigma [34] as well as omega [35] mesons. The model was further extended to investigate the properties of deconfinement [36,38] and nuclear matter [37]. As an alternative approach to address the thermodynamic consistency, an additional fictitious degree of freedom was introduced [41,42] to elaborate a generalized version of the first law of thermodynamics.

From the field theory perspective, the mass of a particle can be defined either by the pole of the effective propagator or via the Debye screen mass extracted at a small momentum, provided the question of the gauge invariance is adequately dealt with. In particular, the calculations with a hard thermal-loop approximation show that the gluon screen mass extracted from the above pictures is consistent [19,48]. The derived quasiparticle mass, in turn, is a function of the temperature and chemical potential. As a result, the above dependence calls for a generalization scheme for thermodynamic consistency. Further developments by Peshier et al. give rise to a flow equation [23–25,49]. The latter is a partial differential equation, and its boundary condition is chosen at a vanishing baryon density, adapted to the lattice QCD data. It was shown that the thermodynamic properties obtained from such a framework agree well with the lattice calculations at the finite baryon chemical potential.

Following [22], one takes the grand partition function of the system $\mathcal{Z}$ as the starting point, which reads

$$\mathcal{Z}(V,T,\mu) = \mathrm{Tr}\Big[e^{-\beta(H-\mu N)}\Big], \tag{1}$$

and β is the reciprocal of the temperature T, μ represents the chemical potential, V is the volume of the system, and H and N are the Hamiltonian and conserved number operators. In order to derive the remaining thermodynamic quantities (such as the pressure, energy density, and conserved number density) in a consistent fashion, the two following identities need to be valid,

$$\frac{\partial \mathcal{Z}(V,T,\mu)}{\partial \beta} = -\mathrm{Tr}\Big[(H-\mu N)e^{-\beta(H-\mu N)}\Big], \tag{2}$$

and

$$\frac{\partial \mathcal{Z}(V,T,\mu)}{\partial \mu} = \beta \mathrm{Tr}\Big[N e^{-\beta(H-\mu N)}\Big]. \tag{3}$$

The conditions Equations (2) and (3) are manifestly satisfied when the Hamiltonian is not medium dependent. As an example, for the quasiparticle model proposed in [17],

$$H = \sum_{\mathbf{k}} \omega(\mathbf{k},T,\mu) a^\dagger_{\mathbf{k}} a_{\mathbf{k}} + VB, \tag{4}$$

where B, the *bag constant*, is attributed to the vacuum energy, mostly viewed as a constant, and

$$\omega(\mathbf{k},T,\mu) = \omega(\mathbf{k},m) = \sqrt{k^2+m^2}, \tag{5}$$

where $k = |\mathbf{k}|$ and $m = m(T)$ is an explicit function of the temperature. The latter adds an additional contribution to the partial derivative in Equation (2), associated with H. The recipe by Gorenstein and Yang is derived from the proposal that B should also be medium dependent, namely $B = B(T)$, whose entire purpose is to identically cancel out the undesirable contribution coming from the temperature-dependent quasiparticle mass. To be explicit, it is not difficult to show [22,23,44] that the above requirement dictates the relation

$$\frac{dB}{dT} = \frac{\partial p_{\mathrm{id}}(T,\mu,m)}{\partial m}\bigg|_{T,\mu} \frac{dm}{dT}, \tag{6}$$

where the pressure of ideal gas is an intensive property given by the standard statistical mechanics,

$$p_{\mathrm{id}} = \frac{T}{V} \ln \mathcal{Z}(V,T,\mu)\bigg|_{B=0}, \tag{7}$$

whose specific form is given below in Equation (13).

Because $m = m(T)$ and $B = B(T)$, we have $B = B(m)$. In other words, Equation (6) implies

$$\frac{dB}{dm} = \frac{\partial p_{\mathrm{id}}(T,\mu,m)}{\partial m}\bigg|_{T,\mu}, \tag{8}$$

where the bag constant B is understood to be a function of the particle mass m only.

Similarly, if the quasiparticle mass is chemical potential dependent, it seems that the above scheme can be readily applied. Specifically, one replaces the temperature derivative in Equation (6) with the chemical potential derivative, while Equation (8) remains unchanged. Moreover, if the mass function depends on both the temperature and chemical potential, namely $m = m(T,\mu)$, Equation (8) seemingly serves the purpose. However, though it might not be apparent at first glimpse, one can argue [44] that Gorenstein and Yang's scheme cannot be applied straightforwardly to such a case. This can be understood as follows. To be precise, one needs to solve for $B = B(m)$ for an arbitrarily given form

$m = m(T,\mu)$, using Equation (8). Observing the l.h.s. of Equation (8), one concludes that the dependence of the temperature and chemical potential can be entirely "packed" into the quasiparticle mass m. On the other hand, because the form $m = m(T,\mu)$ is arbitrary, one can always redefine this function so that the r.h.s. of Equation (8) cannot be written as a function of m.

We note that the above considerations do not necessarily invalidate Equation (8). Instead, it indicates the existence of some additional constraint when finite chemical potential is involved. In [23], Peshier et al. derived a flow equation giving a further constraint for the mass function $m = m(T,\mu)$. In [44], some of us derived an integro-differential equation, which is shown to fall back to the flow equation under certain circumstances. Moreover, it was demonstrated that there are also other possibilities, and in particular, the quasiparticle mass can be a function of the momentum.

In the present study, we proceed to further explore the topic by adopting a "bottom-up" approach. Specifically, instead of numerically adjusting the model parameters to the lattice QCD data, we choose a straightforward but analytical form for the mass function at the vanishing chemical potential. By adopting the analytic function, one can scrutinize the different branches of the mass function in the temperature–chemical potential parameter plane. The remainder of the present paper is organized as follows. In the next section, we review the relevant elements regarding the thermodynamic consistency in the quasiparticle model. The resulting integro-differential equation is presented and discussed. Section 3 focuses on the novel type of solution. In particular, we explore a mathematically simple form of the mass function at the vanishing chemical potential. It is shown that such a choice will not entirely determine the mass function in the temperature–chemical potential parameter plane. Different possibilities are then investigated numerically. The last section is devoted to further discussions and concluding remarks.

2. The Generalized Condition for Thermodynamical Consistency

This section discusses the formal constraints for the thermodynamic consistency in the quasiparticle model. For the present study, the term *consistency* implies the following three essential aspects. First, all the thermodynamic quantities can be derived using the standard formulae once the grand partition function $\mathcal{Z}$ is given. Second, these thermodynamic quantities possess an interpretation in accordance with the ensemble average in statistics. Last but not least, most thermodynamic identities, for instance, those based on the first law of thermodynamics (c.f. Equation (19)) and extensive properties (c.f. Equation (17)), remain unchanged. To our knowledge, the scheme proposed by Gorenstein and Yang is the only one that meets all three above requirements.

As discussed in [22], once Equations (2) and (3) are satisfied, the energy density and particle number density derived either from the ensemble average or from the partial derivative of the grand partition function possess identical forms. These lead to the following forms of the energy density

$$\varepsilon = \frac{\langle E\rangle}{V} = -\frac{1}{V}\frac{\partial \ln \mathcal{Z}}{\partial \beta} = \epsilon_{\rm id} + B, \tag{9}$$

with

$$\epsilon_{\rm id} = \frac{g}{2\pi^2}\int_0^\infty \frac{k^2 dk\omega(k,T,\mu)}{\exp[(\omega(k,T,\mu)-\mu)/T]\mp 1} + \text{c.t.}\,, \tag{10}$$

where g indicates possible degeneracy, "$\mp$" corresponds to boson and fermion, and the counter term "c.t." indicates contributions from anti-particles obtained by the substitution $\mu \to -\mu$ in the foregoing term. We have also considered the isotropic case $m(\mathbf{k},T,\mu) = m(k,T,\mu)$.

To derive the above equation, we have assumed the validity of Equation (2), namely the contribution from the temperature dependence of quasiparticle mass has precisely been canceled out with the temperature dependence of B. By writing it out explicitly, one finds

$$\frac{\partial B}{\partial T} = -\frac{g}{2\pi^2}\int_0^\infty \frac{k^2dk}{\omega(k,T,\mu)}\frac{1}{\exp[(\omega(k,T,\mu)-\mu)/T]\mp 1}m\frac{\partial m}{\partial T} + \text{c.t.} \tag{11}$$

In statistical mechanics, the pressure is interpreted as a "general force", which reads

$$p = \frac{1}{\beta}\frac{\partial \ln \mathcal{Z}}{\partial V} = \frac{1}{\beta}\frac{\ln \mathcal{Z}}{V} = p_{\text{id}} - B, \tag{12}$$

where

$$\begin{aligned} p_{\text{id}} &= \frac{\mp g}{2\pi^2}\int_0^\infty k^2 dk \ln\{1\mp\exp[(\mu-\omega(k,T,\mu))/T]\} + \text{c.t.} \\ &= \frac{g}{12\pi^2}\int_0^\infty \frac{k^3dk}{\exp[(\omega(k,T,\mu)-\mu)/T]\mp 1}\left.\frac{\partial\omega(k,T,\mu)}{\partial k}\right|_{T,\mu} + \text{c.t.} \end{aligned} \tag{13}$$

Moreover, as an ensemble average, the number density is found to be

$$n = \frac{\langle N\rangle}{V} = -\frac{1}{V}\frac{\partial \ln \mathcal{Z}}{\partial \alpha} = n_{\text{id}}, \tag{14}$$

where

$$n_{\text{id}} = \frac{g}{2\pi^2}\int_0^\infty \frac{k^2dk}{\exp[(\omega(k,T,\mu)-\mu)/T]\mp 1} - \text{c.t.} \tag{15}$$

Again, we have assumed the condition Equation (3), which states that the contribution from the chemical potential dependence of quasiparticle mass in the ideal gas term and that from the bag constant B cancel out each other. The above condition can be specified to give

$$\frac{\partial B}{\partial \mu} = -\frac{g}{2\pi^2}\int_0^\infty \frac{k^2dk}{\omega(k,T,\mu)}\frac{1}{\exp[(\omega(k,T,\mu)-\mu)/T]\mp 1}m\frac{\partial m}{\partial \mu} + \text{c.t.} \tag{16}$$

The well-known thermodynamic identity,

$$\epsilon = T\frac{\partial p}{\partial T} - p + \mu n, \tag{17}$$

essentially comes from the first law of thermodynamics and its extensive properties. As a matter of fact, following the procedure elaborated in the standard textbook [50], it is not difficult to verify that the total derivative of $q = \ln \mathcal{Z}$ gives

$$dq = -\langle N\rangle d\alpha - \langle E\rangle d\beta - \beta p dV. \tag{18}$$

By comparing the above expression with the first law of thermodynamics, namely

$$d\langle E\rangle = TdS - pdV + \mu d\langle N\rangle, \tag{19}$$

we have the mapping

$$\beta = \frac{1}{k_B T},$$
$$\alpha = -\frac{\mu}{k_B T},$$
$$q + \alpha N + \beta E = \frac{S}{k_B}. \tag{20}$$

The validity of Equation (17) is readily verified.

Now, we proceed to discuss the implications of the conditions Equations (11) and (16). By taking the partial derivative of Equation (11) in μ and comparing with the partial derivative of Equation (16) in T, one arrives at the following integro-differential equation [44]

$$\langle\!\langle m \frac{\partial m}{\partial T} \rangle\!\rangle_- = \langle\!\langle m \frac{\partial m}{\partial \mu} \rangle\!\rangle_+, \tag{21}$$

where

$$\begin{aligned} \langle\!\langle O \rangle\!\rangle_- &\equiv \int_0^\infty k^2 dk \left\{ \frac{\exp[(\omega-\mu)/T]}{(\exp[(\omega-\mu)/T] \mp 1)^2 \omega T} - \text{c.t.} \right\} O(k), \\ \langle\!\langle O \rangle\!\rangle_+ &\equiv \int_0^\infty k^2 dk \left\{ \frac{\exp[(\omega-\mu)/T](\omega-\mu)}{(\exp[(\omega-\mu)/T] \mp 1)^2 \omega T^2} + \text{c.t.} \right\} O(k). \end{aligned} \tag{22}$$

The solution of Equation (21), $m = m(k, T, \mu)$, is in general a function also of the momentum k. In turn, the bag constant B is obtained by integrating Equations (11) and (16) on the parameter plane. It can be viewed as a functional of $m(k, T, \mu)$ besides being a function of T and μ. It is noted that the above discussions can be straightforwardly generalized to the case where the system is not isotropic, where $m = m(\mathbf{k}, T, \mu)$. As pointed out in [44], if one simplifies and considers the momentum-independent case, namely $m(\mathbf{k}, T, \mu) = m(T, \mu)$, one readily falls back to the flow equation derived in Ref. [23]. In this case, B also simplifies to a function of T and μ. We are, however, more interested in exploring the momentum-dependent case, which will be elaborated further in the following section.

3. Bottom-Up Toy Model Approaches

An apparent momentum-dependent solution to Equation (21) can be obtained by "factoring out" the momentum integration $\int k^2 dk$ and assuming the integrand vanishes. In other words,

$$\left\{ \frac{\exp[(\omega-\mu)/T]T}{(\exp[(\omega-\mu)/T] \mp 1)^2} - \text{c.t.} \right\} \frac{\partial m}{\partial T} = \left\{ \frac{\exp[(\omega-\mu)/T](\omega-\mu)}{(\exp[(\omega-\mu)/T] \mp 1)^2} + \text{c.t.} \right\} \frac{\partial m}{\partial \mu}. \tag{23}$$

The above equation can be solved by using the method of characteristics [51]. To be specific, for a given k, the solution is the surface tangent to the vector field

$$(a(T, \mu, m), b(T, \mu, m), 0),$$

where

$$\begin{aligned} a(T, \mu, m) &= \frac{\exp[(\omega-\mu)/T]T}{(\exp[(\omega-\mu)/T] \mp 1)^2} - \text{c.t.}, \\ b(T, \mu, m) &= -\frac{\exp[(\omega-\mu)/T](\omega-\mu)}{(\exp[(\omega-\mu)/T] \mp 1)^2} - \text{c.t.}. \end{aligned} \tag{24}$$

Its formal solution is the characteristic curves obtained by the integration

$$\begin{aligned} \frac{dT}{d\lambda} &= a(T, \mu, m), \\ \frac{d\mu}{d\lambda} &= b(T, \mu, m), \end{aligned} \tag{25}$$

where λ is an intermediate variable, for given k, m, and thus ω.

An interesting scenario that gives rise to an analytic solution occurs when one ignores the anti-particles' contributions. By taking ω, T, μ as the three independent variables, the method of characteristics gives [44] the following formal solution

$$m = f\left(\frac{T\omega}{\omega - \mu}, k\right), \tag{26}$$

where an arbitrary function f furnishes the boundary condition at vanishing chemical potential, namely $f(T) \equiv f(T, 0) = m(T, \mu = 0, k = 0)$, where we assume that the mass is independent of the momentum at $\mu = 0$. We note that the resultant mass function is a function of k, T, and m, and therefore the solution of the form Equation (26) serves as a simple but non-trivial example.

In [44], the freedom in f was employed to perform a numerical fit to the lattice QCD results for the $N_f = 2 + 1$ flavor QCD system [52–56] at the vanishing chemical potential. Then, the relevant physical quantities, such as the trace anomaly, sound velocity, and particle number susceptibility, were evaluated and compared to the lattice data. Instead, we adopt a "bottom-up" approach for the present study. Specifically, we consider two cases where one assumes a simple ansatz for f posteriorly adapted to the lattice results and proceeds analytically to a large extent.

Case 1: Our first choice is a simple linear fit. Based on the lattice data [52] shown in Figure 1, there are two regions where the mass of the quasiparticle is primarily a linear function in temperature. In other words,

$$f|_{\mu=0} = c_1 T + c_2, \tag{27}$$

which gives

$$f\left(\frac{T\omega}{\omega - \mu}, k\right) = f\left(\frac{T\omega}{\omega - \mu}\right) = \frac{c_1 T\omega}{\omega - \mu} + c_2. \tag{28}$$

Despite its simple form, Equation (27) might be plagued by the pole on its denominator. To avoid the pole at $\omega = \mu$ for an arbitrary momentum k indicates the condition

$$\omega > \mu, \tag{29}$$

that is, $\omega \ge m > \mu$, by considering the definition Equation (5). Otherwise, if one requires $\omega < \mu$, it is always possible to find a momentum k large enough to violate the condition.

Substituting Equation (28) into Equation (26) gives

$$\omega(m - c_1 T - c_2) = \mu(m - c_2), \tag{30}$$

for which Equation (29) dictates

$$c_1 > 0, \tag{31}$$

while given $T > 0$ and $\mu > 0$.

By substituting Equation (5) into Equation (30) and squaring both sides, one finds a fourth-degree polynomial equation for m. This equation possesses four roots, where complex roots always appear in pairs. The physically relevant solution must sit on the positive real axis.

From this point on, we proceed numerically. One extracts the values c_1 and c_2 from the region shown in Figure 1, satisfying $c_1 > 0$, and finds $c_1 = 0.44$ and $c_2 = 0.13$. The fourth-degree polynomial contains a pair of complex roots, which are subsequently discarded. One of the remaining two real roots is extraneous, owing to the fact that we have squared both sides of Equation (30). The resultant mass function $m(k, T, \mu)$ and the bag constant

$B(T, \mu)$ are shown in Figure 2. The left and middle plots give the mass as a function of T and μ at given $k = 1$ GeV and that of k and μ at given $T = 0.25$ GeV. The resulting bag constant is obtained by numerical integration of Equations (11) and (16). The dependence of the bag constant on the temperature T or chemical potential μ is presented in the right plot of Figure 2. The mass function and the bag constant are found to be moderate in T and μ. As $k \to 0$, according to the middle plot of Figure 2, the quasiparticle mass increases significantly. It is noted that the obtained bag constant B is manifestly path independent. For instance, one evaluates $B(T, \mu)$ by using two following integration paths on the $T - \mu$ plane. The integration for B is carried out from ($T_0 = 0.25$ GeV , $\mu_0 = 0$) to ($T_1 = 0.45, \mu_1 = 0.3$), where path 1 is defined by $(T_0, \mu_0) \to (T_1, \mu_0) \to (T_1, \mu_1)$, while path 2 is through $(T_0, \mu_0) \to (T_0, \mu_1) \to (T_1, \mu_1)$. One finds

$$[B(T_1, \mu_1) - B(T_0, \mu_0)]|_{\text{path 1}} = -0.606049 = [B(T_1, \mu_1) - B(T_0, \mu_0)]|_{\text{path 2}}.$$

We also note that based on the above discussions, the fit to the region $c_1 < 0$, where the mass of the quasiparticle decreases with increasing temperature in Figure 1, is doomed to fail. A numerical attempt reveals path-dependent values, which signals that those obtained by straightforward integration do not yield mathematically well-defined results. This is due to the undesirable pole at $\omega = \mu$ in the denominator of the first term on the r.h.s. of Equation (28). In order to handle the region where $c_1 < 0$, we proceed to consider the second case.

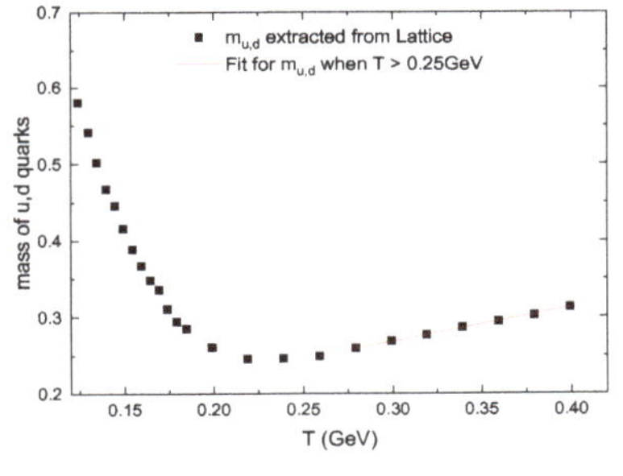

Figure 1. The mass of u and d quarks at vanishing chemical potential derived from the lattice data [52]. It can be readily extracted for vanishing chemical potential using Gorenstein and Yang's scheme [22]. The curve is then fit to analytic form Equation (27) discussed in the text.

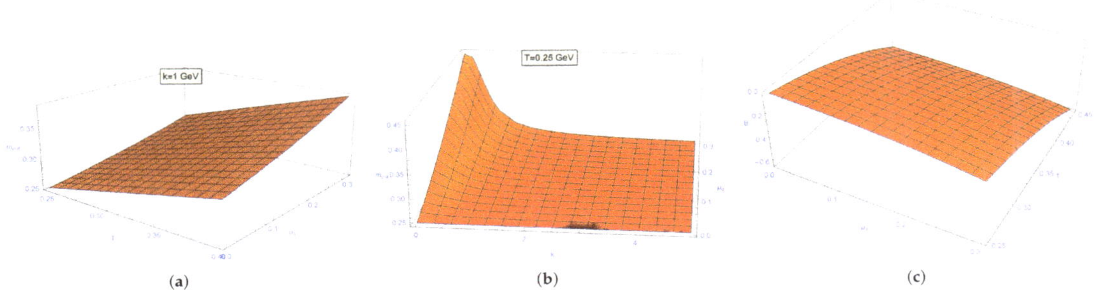

(a) (b) (c)

Figure 2. The derived quasiparticle mass in the parameter space according to the form given by Equation (28) and the fit shown in Figure 1. (**a**) The quasiparticle mass m as a function of T and μ at $k = 1$ GeV. (**b**) The quasiparticle mass m as a function of k and μ at $T = 0.25$ GeV. (**c**) The bag constant B as a function of T and μ.

Case 2: The second choice involves a linear function in the reciprocal of the argument of Equation (26). To be specific, we consider the ansatz,

$$f\big|_{\mu=0} = \frac{1}{c_3 + c_4 T}, \tag{32}$$

which gives

$$f\left(\frac{T\omega}{\omega-\mu}, k\right) = \frac{\omega-\mu}{c_3(\omega-\mu) + c_4 T\omega}. \tag{33}$$

In this case, to avoid the pole in the denominator, one considers the following constraint

$$\omega(c_4 T + c_3) > \mu c_3. \tag{34}$$

Substituting Equation (33) into Equation (26) gives

$$\omega(c_4 T m + m c_3 - 1) = \mu(m c_3 - 1). \tag{35}$$

Given $T > 0$ and $m > 0$, Equation (34) implies

$$\omega > \mu, \;\; c_4 > 0, \tag{36}$$

and moreover, Equation (35) further indicates

$$c_3 < \frac{1}{m}, \;\; c_4 T m + m c_3 < 1. \tag{37}$$

Otherwise, if $c_3 \geq 1/m$, Equation (35) can no longer hold.

To proceed, one substitutes Equation (5) into Equation (35) and squares both sides, and one again finds a fourth-degree polynomial equation for m. Similarly, this equation possesses four roots, where complex roots appear in pairs.

We proceed numerically from this point on. The values c_3 and c_4 are extracted from a fit to the lattice QCD data shown in Figure 3. One finds $c_3 = -1.84$ and $c_4 = 28.55$, which affirms the second choice above. By discussing a pair of complex roots and an extraneous root, the physically relevant solution is eventually singled out from the two sitting on the positive real axis.

The resultant mass function $m(k, T, \mu)$ and the bag constant $B(T, \mu)$ are shown in Figure 4. The left and middle plots give the mass as a function of T and μ at given $k = 1$ and as a function of k and μ at given $T = 0.12$. Again, the bag constant can be obtained by the numerical integration of Equations (11) and (16). The dependence of the bag constant on the temperature T or chemical potential μ is presented in the right plot of Figure 4. The mass function and the bag constant are found to be moderate in T and μ, mainly in accordance with the existing results [44]. Different from Figure 2, as $k \to 0$, the mass of the quasiparticle does not modify significantly. Again, the obtained bag constant B is manifestly path independent.

Before closing this section, we present in Figure 5 a few resulting thermodynamic quantities evaluated using the toy model proposed in case 2. In the left plot, we show the pressure, energy density, and entropy density as a function of the temperature at the vanishing chemical potential. The right plot gives the difference in the pressure between the states with the finite and vanishing chemical potential. It is noted in the calculations, in accordance with the simplified scenario, one only takes into account the u and d quarks but does not include the s quarks, gluons, or anti-particles. By comparing the results with those obtained using more sophisticated approaches [26,44,57], one is led to the following observations. The tendency of the temperature dependence is mainly correct, while the magnitudes of the calculated thermodynamic quantities consistently underestimate the existing results. This is because the simplified models do not consider the contributions from the remaining degrees of freedom, including those from the anti-particles. Moreover, the order of magnitude for these quantities can be roughly recuperated by multiplying a factor of two. The latter effectively compensates for the contributions missing from the anti-particles. We note, nonetheless, that the main objective of the present approach is to explore the analytic properties of the mass function from a bottom-up perspective rather than reproduce the lattice data numerically by employing some sophisticated approximate function.

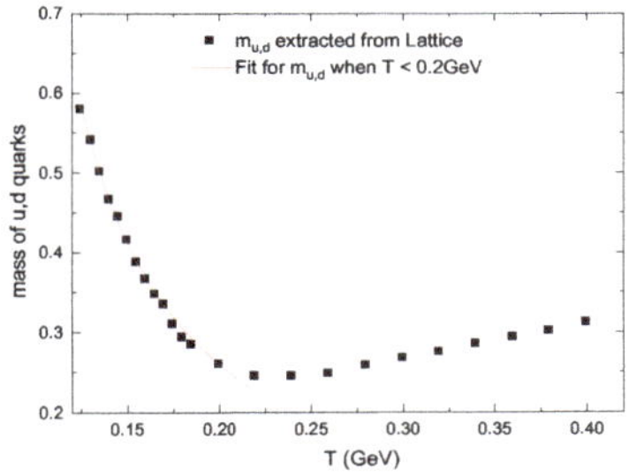

Figure 3. The mass of u and d quarks at vanishing chemical potential derived from the lattice data [52], which is fit to analytic form Equation (32) discussed in the text.

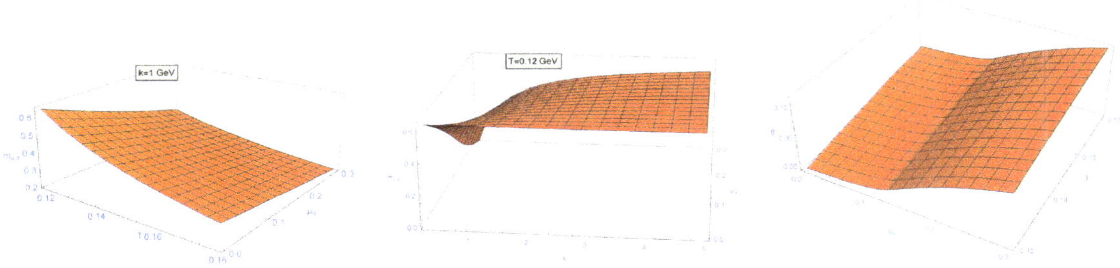

(**a**) (**b**) (**c**)

Figure 4. The derived quasiparticle mass in the parameter space according to the form given by Equation (33) and the fit shown in Figure 3. (**a**) The quasiparticle mass m as a function of T and μ at $k = 1$ GeV. (**b**) The quasiparticle mass m as a function of k and μ at $T = 0.12$ GeV. (**c**) The bag constant B as a function of T and μ.

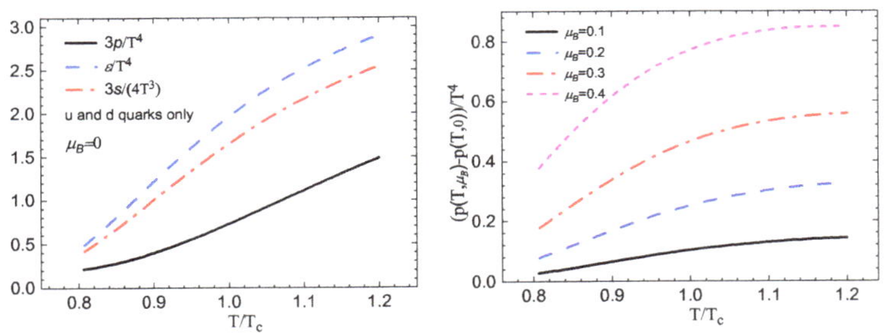

(**a**) (**b**)

Figure 5. The derived thermodynamic quantities by considering the u and d quarks, where the s quarks, gluons, and anti-particles are not explicitly taken into account. (**a**) The pressure $3p/T^4$, energy density ϵ/T^4, and entropy density $3s/(4T^3)$ as functions of the temperature T/T_c at vanishing chemical potential, where $T_c = 1.5$ GeV in accordance with lattice QCD data. (**b**) The difference in pressure between the states with finite and vanishing chemical potential is shown as a function of temperature T/T_c.

4. Concluding Remarks

To summarize, in this work, we reviewed the topic of the quasiparticle model closely related to Ru-Keng Su's distinguished contributions in the past years. Moreover, we explore the approach applied to scenarios with finite chemical potential. Different from the standard recipe in the literature, we explored the possibility that the effective mass of the quasiparticle might be a function of the momentum, in addition to the dependence on the temperature and chemical potential. It was shown that such a scenario emerges as a special solution to an integro-differential equation derived from the thermodynamic consistency. We pointed out that the special solution in question is essentially a generalization to those previously explored in the literature. Instead of fitting to the lattice QCD data at the vanishing chemical potential, we performed a "bottom-up" approach by assuming two analytic ansatzes. The remaining physical quantities were subsequently derived and discussed. We note that the momentum-dependent quanta mass was also addressed by some authors from the QCD perspective, where the analyses were closely related to the symmetry of the underlying system. In terms of the Gribov–Zwanziger framework, the results on the gluon [58–61] and quark propagator [62] indicated that the pole masses are functions of the momentum. In addition, calculations using the Schwinger–Dyson equation showed the momentum dependence for both gluon [63] and quark [64,65] dynamic masses.

The current approach's main objective is to explore the analytic properties of the mass function. It is primarily motivated as one might distinguish the various roots deriving from the thermodynamical consistency condition. As observed and discussed in the main text, these different roots are somehow separated by the pole of the relevant equation, which is not apparent if a numerical scheme was utilized in the first place. The calculations primarily employ Equation (26). It is a simplified approach as it ignores the anti-particles' contributions and is only utilized to fit to accommodate the u and d quarks. On the other hand, a numerical approach directly based on Equation (21) was carried out in a previous study [44], where the cancelations warranted by Equations (11) and (16) take place for individual particles, as well as their anti-particles. Nonetheless, the present study gives rise to the following speculations. First, we have attempted to avoid the singularity of the mass function by entirely evading its poles by imposing the conditions, Equations (29) and (34). The resultant physical quantities are, in turn, manifestly *analytic* on the T and μ parameter space. Curiously, from a theoretical perspective, one expects a curve of a first-order phase transition on the parameter plane, which entails some discontinuity. In other words, the discontinuity avoided in the present study might be utilized in our favor. Specifically, a pole in the mass function indicates an infinite mass, which can be viewed as a natural and benign outcome when a degree of freedom can hardly be excited. We plan to address these aspects in further studies.

Author Contributions: Conceptualization, methodology, investigation and writing: all the authors; software: H.-H.M.; validation: H.-H.M. and W.-L.Q.; All authors have read and agreed to the published version of the manuscript.

Funding: This research is partly funded by the National Natural Science Foundation of China (No. 12047506) the Central Government Guidance Funds for Local Scientific and Technological Development, China (No. Guike ZY22096024). It is also supported by Brazilian agencies Fundação de Amparo à Pesquisa do Estado de São Paulo, Conselho Nacional de Desenvolvimento Científico e Tecnológico, and Coordenação de Aperfeiçoamento de Pessoal de Nível Superior.

Data Availability Statement: The datasets generated and analyzed during the current study are available from the corresponding author upon reasonable request.

Conflicts of Interest: The authors declare no conflict of interest.

References

1. Karsch, F. Lattice QCD at high temperature and density. *Lect. Notes Phys.* **2002**, *583*, 209.
2. Hwa, R.C.; Wang, X.N. (Eds.) *Quark-Gluon Plasma 3*; World Scientific: Singapore, 2004.

3. Florkowski, W.; Heller, M.P.; Spalinski, M. New theories of relativistic hydrodynamics in the LHC era. *Rept. Prog. Phys.* **2018**, *81*, 046001. [CrossRef] [PubMed]
4. Hama, Y.; Kodama, T.; Socolowski, O., Jr. Topics on hydrodynamic model of nucleus-nucleus collisions. *Braz. J. Phys.* **2005**, *35*, 24. [CrossRef]
5. de Souza, R.D.; Koide, T.; Kodama, T. Hydrodynamic Approaches in Relativistic Heavy Ion Reactions. *Prog. Part. Nucl. Phys.* **2016**, *86*, 35. [CrossRef]
6. Hama, Y.; Kodama, T.; Qian, W.-L. Two-particle correlations at high-energy nuclear collisions, peripheral-tube model revisited. *J. Phys.* **2021**, *G48*, 015104. [CrossRef]
7. Qian, W.-L.; Andrade, R.; Socolowsk, O., Jr.; Grassi, F.; Kodama, T.; Hama, Y. p(T) distribution of hyperons in 200-A-GeV Au-Au in smoothed particle hydrodynamics. *Braz. J. Phys.* **2007**, *37*, 767. [CrossRef]
8. Qian, W.L.; Andrade, R.; Grassi, F.; Socolowski, O., Jr.; Kodama, T.; Hama, Y. Effect of chemical freeze out on identified particle spectra at 200-A-GeV Au-Au Collisions at RHIC using SPheRIO. *Int. J. Mod. Phys.* **2007**, *E16*, 1877. [CrossRef]
9. Hama, Y.; Andrade, R.P.G.; Grassi, F.; Qian, W.L.; Kodama, T. Fluctuation of the Initial Conditions and Its Consequences on Some Observables. *Acta Phys. Polon. B* **2009**, *40*, 931.
10. Andrade, R.; Grassi, F.; Hama, Y.; Qian, W.-L. A Closer look at the influence of tubular initial conditions on two-particle correlations. *J. Phys. G* **2010**, *G37*, 094043. [CrossRef]
11. Andrade, R.P.G.; Grassi, F.; Hama, Y.; Qian, W.-L. Temporal evolution of tubular initial conditions and their influence on two-particle correlations in relativistic nuclear collisions. *Phys. Lett.* **2012**, *B712*, 226. [CrossRef]
12. Andrade, R.; Grassi, F.; Hama, Y.; Qian, W.-L. Hydrodynamics: Fluctuating Initial Conditions and Two-particle Correlations. *Nucl. Phys.* **2011**, *A854*, 81. [CrossRef]
13. Castilho, W.M.; Qian, W.-L.; Gardim, F.G.; Hama, Y.; Kodama, T. Hydrodynamic approach to the centrality dependence of di-hadron correlations. *Phys. Rev.* **2017**, *C95*, 064908. [CrossRef]
14. Castilho, W.M.; Qian, W.-L.; Hama, Y.; Kodama, T. Event-plane dependent di-hadron correlations with harmonic v_n subtraction in a hydrodynamic model. *Phys. Lett.* **2018**, *B777*, 369. [CrossRef]
15. Wen, D.; Lin, K.; Qian, W.L.; Wang, B.; Hama, Y.; Kodama, T. On nonlinearity in hydrodynamic response to the initial geometry in relativistic heavy-ion collisions. *Eur. Phys. J.* **2020**, *A56*, 222. [CrossRef]
16. Fowler, G.; Raha, S.; Weiner, R. CONFINEMENT AND PHASE TRANSITIONS. *Z. Phys.* **1981**, *C9*, 271. [CrossRef]
17. Peshier, A.; Kampfer, B.; Pavlenko, O.P.; Soff, G. An Effective model of the quark - gluon plasma with thermal parton masses. *Phys. Lett.* **1994**, *B337*, 235. [CrossRef]
18. Ginsparg, P.H. First Order and Second Order Phase Transitions in Gauge Theories at Finite Temperature. *Nucl. Phys.* **1980**, *B170*, 388. [CrossRef]
19. Pisarski, R.D. Renormalized Gauge Propagator in Hot Gauge Theories. *Physica* **1989**, *A158*, 146. [CrossRef]
20. Pisarski, R.D. Quark gluon plasma as a condensate of SU(3) Wilson lines. *Phys. Rev.* **2000**, *D62*, 111501.
21. Asakawa, M.; Ko, C.M. Seeing the QCD phase transition with phi mesons. *Phys. Lett.* **1994**, *B322*, 33. [CrossRef]
22. Gorenstein, M.I.; Yang, S.-N. Gluon plasma with a medium dependent dispersion relation. *Phys. Rev.* **1995**, *D52*, 5206. [CrossRef] [PubMed]
23. Peshier, A.; Kampfer, B.; Soff, G. The Equation of state of deconfined matter at finite chemical potential in a quasiparticle description. *Phys. Rev.* **2000**, *C61*, 045203. [CrossRef]
24. Peshier, A.; Kampfer, B.; Soff, G. From QCD lattice calculations to the equation of state of quark matter. *Phys. Rev.* **2002**, *D66*, 094003. [CrossRef]
25. Bluhm, M.; Kampfer, B.; Soff, G. The QCD equation of state near T(c) within a quasi-particle model. *Phys. Lett.* **2005**, *B620*, 131. [CrossRef]
26. Biro, T.S.; Shanenko, A.A.; Toneev, V.D. Towards thermodynamical consistency of quasiparticle picture. *Phys. Atom. Nucl.* **2003**, *66*, 982. [CrossRef]
27. Bannur, V.M. Comments on quasiparticle models of quark-gluon plasma. *Phys. Lett.* **2007**, *B647*, 271. [CrossRef]
28. Wang, P., Strange matter in a selfconsistent quark mass density dependent model. *Phys. Rev. C* **2000**, *62*, 015204. [CrossRef]
29. Zhang, Y.; Su, R.-K.; Ying, S.-Q.; Wang, P. Quark mass density and temperature dependent model for strange quark matter. *EPL* **2001**, *53*, 361. [CrossRef]
30. Zhang, Y.; Su, R.-K. Quark mass density and temperature dependent model for bulk strange quark matter. *Phys. Rev. C* **2002**, *65*, 035202. [CrossRef]
31. Zhang, Y.; Su, R.-K. Stability of strangelet at finite temperature. *Phys. Rev. C* **2003**, *67*, 015202. [CrossRef]
32. Zhang, Y.; Su, R.-K. Dibaryon systems in the quark mass density and temperature dependent model. *J. Phys. G* **2004**, *30*, 811. [CrossRef]
33. Qian, W.L.; Su, R.-K.; Song, H.Q. Warm strange hadronic matter in an effective model with a weak Y-Y interaction. *J. Phys.* **2004**, *G30*, 1893. [CrossRef]
34. Wu, C.; Qian, W.L.; Su, R.-K. Improved quark mass density- dependent model with quark and non-linear scalar field coupling. *Phys. Rev. C* **2005**, *72*, 035205. [CrossRef]
35. Wu, C.; Qian, W.-L.; Su, R.-K. Improved quark mass density- dependent model with quark-sigma meson and quark-omega meson couplings. *Phys. Rev. C* **2008**, *77*, 015203. [CrossRef]

36. Wu, C.; Su, R.-K. Quark deconfinement phase transition for improved quark mass density-dependent model. *J. Phys. G* **2008**, *35*, 125001. [CrossRef]
37. Wu, C.; Su, R.-K. Nuclear matter and neutron matter for improved quark mass density- dependent model with rho mesons. *J. Phys. G* **2009**, *36*, 095101. [CrossRef]
38. Mao, H.; Su, R.-K.; Zhao, W.-Q. Soliton solutions of the improved quark mass density-dependent model at finite temperature. *Phys. Rev.* **2006**, *C74*, 055204. [CrossRef]
39. Yin, S.; Su, R.-K. Thermodynamics of system with density- and/or temperature-dependent mass particles. *arXiv* **2007**, arXiv:0704.3689.
40. Yin, S.; Su, R.-K. The Intrinsic degree of freedom for quasiparticle in thermodynamics with medium effects. *arXiv* **2007**, arXiv:0709.0179.
41. Yin, S.; Su, R.-K. Consistent thermodynamic treatment for a quark-mass density-dependent model. *Phys. Rev.* **2008**, *C77*, 055204. [CrossRef]
42. Yin, S.; Su, R.-K. Consistent Thermodynamics for Quasiparticle Boson System with Zero Chemical Potential. *Int. J. Mod. Phys.* **2010**, *A25*, 1449. [CrossRef]
43. Ma, H.-H.; Qian, W.-L. A quasiparticle equation of state with a phenomenological critical point. *Braz. J. Phys.* **2018**, *48*, 160. [CrossRef]
44. Ma, H.-H.; Lin, K.; Qian, W.-L.; Hama, Y.; Kodama, T. Thermodynamical consistency of quasiparticle model at finite baryon density. *Phys. Rev.* **2019**, *C100*, 015206. [CrossRef]
45. Friedberg, R.; Lee, T.D. Fermion Field Nontopological Solitons. 1. *Phys. Rev. D* **1977**, *15*, 1694. [CrossRef]
46. Friedberg, R.; Lee, T.D. Fermion Field Nontopological Solitons. 2. Models for Hadrons. *Phys. Rev. D* **1977**, *16*, 1096. [CrossRef]
47. Goldflam, R.; Wilets, L. The Soliton Bag Model. *Phys. Rev. D* **1982**, *25*, 1951. [CrossRef]
48. Flechsig, F.; Rebhan, A.K.; Schulz, H. The Infrared sensitivity of screening and damping in a quark-gluon plasma. *Phys. Rev.* **1995**, *D52*, 2994. [CrossRef]
49. Bluhm, M.; Kampfer, B.; Schulze, R.; Seipt, D.; Heinz, U. A family of equations of state based on lattice QCD: Impact on flow in ultrarelativistic heavy-ion collisions. *Phys. Rev.* **2007**, *C76*, 034901. [CrossRef]
50. Pathria, R.K. *Statistical Mechanics*; Butterworth-Heinemann: Oxford, UK, 1996.
51. Courant, R.; Hilbert, D. *Methods of Mathematical Physics*, 1st ed.; Wiley-Interscience: Hoboken, NJ, USA, 1989.
52. Borsanyi, S.; Fodor, Z.; Katz, S.D.; Krieg, S.; Ratti, C.; Szabó, K. Fluctuations of conserved charges at finite temperature from lattice QCD. *J. High Energy Phys.* **2012**, *1*, 138. [CrossRef]
53. Borsányi, S.; Fodor, Z.; Hoelbling, C.; Katz, S.D.; Krieg, S.; Szabo, K.K. Full result for the QCD equation of state with 2+1 flavors. *Phys. Lett.* **2014**, *B730*, 99. [CrossRef]
54. Bazavov, A.; Bhattacharya, T.; DeTar, C.E.; Ding, H.T.; Gottlieb, S.; Gupta, R.; Hegde, P.; Heller, U.M.; Karsch, F.; Laermann, E.; et al. Fluctuations and Correlations of net baryon number, electric charge, and strangeness: A comparison of lattice QCD results with the hadron resonance gas model. *Phys. Rev.* **2012**, *D86*, 034509. [CrossRef]
55. Katz, S.D. Equation of state from lattice QCD. *Nucl. Phys.* **2006**, *A774*, 159. [CrossRef]
56. Bazavov, A.; Ding, H.T.; Hegde, P.; Kaczmarek, O.; Karsch, F.; Laermann, E.; Maezawa, Y.; Mukherjee, S.; Ohno, H.; Petreczky, P.; et al. The QCD Equation of State to $\mathcal{O}(\mu_B^6)$ from Lattice QCD. *Phys. Rev.* **2017**, *D95*, 054504. [CrossRef]
57. Bluhm, M.; Kampfer, B.; Schulze, R.; Seipt, D. Quasi-Particle Description of Strongly Interacting Matter: Towards a Foundation. *Eur. Phys. J.* **2007**, *C49*, 205. [CrossRef]
58. Dudal, D.; Gracey, J.A.; Sorella, S.P.; Vandersickel, N.; Verschelde, H. A Refinement of the Gribov-Zwanziger approach in the Landau gauge: Infrared propagators in harmony with the lattice results. *Phys. Rev.* **2008**, *D78*, 065047. [CrossRef]
59. Dudal, D.; Sorella, S.P.; Vandersickel, N. The dynamical origin of the refinement of the Gribov-Zwanziger theory. *Phys. Rev.* **2011**, *D84*, 065039. [CrossRef]
60. Cucchieri, A.; Dudal, D.; Mendes, T.; Vandersickel, N. Modeling the Gluon Propagator in Landau Gauge: Lattice Estimates of Pole Masses and Dimension-Two Condensates. *Phys. Rev.* **2012**, *D85*, 094513. [CrossRef]
61. Capri, M.A.; Dudal, D.; Pereira, A.D.; Fiorentini, D.; Guimaraes, M.S.; Mintz, B.W.; Palhares, L.F.; Sorella, S.P. Nonperturbative aspects of Euclidean Yang-Mills theories in linear covariant gauges: Nielsen identities and a BRST-invariant two-point correlation function. *Phys. Rev.* **2017**, *D95*, 045011. [CrossRef]
62. Bandyopadhyay, A.; Haque, N.; Mustafa, M.G.; Strickland, M. Dilepton rate and quark number susceptibility with the Gribov action. *Phys. Rev.* **2016**, *D93*, 065004. [CrossRef]
63. Aguilar, A.C.; Binosi, D.; Papavassiliou, J. The Gluon Mass Generation Mechanism: A Concise Primer. *Front. Phys.* **2016**, *11*, 111203. [CrossRef]
64. Aguilar, A.C.; Papavassiliou, J. Chiral symmetry breaking with lattice propagators. *Phys. Rev.* **2011**, *D83*, 014013. [CrossRef]
65. Aguilar, A.C.; Cardona, J.C.; Ferreira, M.N.; Papavassiliou, J. Quark gap equation with non-abelian Ball-Chiu vertex. *arXiv* **2018**, arXiv:1804.04229.

symmetry

MDPI

Review

Heavy-Flavour Jets in High-Energy Nuclear Collisions

Sa Wang [1,2], Wei Dai [3], Enke Wang [1,2], Xin-Nian Wang [4] and Ben-Wei Zhang [1,2,*]

[1] Guangdong Provincial Key Laboratory of Nuclear Science, Institute of Quantum Matter, South China Normal University, Guangzhou 510006, China
[2] Key Laboratory of Quark & Lepton Physics (MOE) and Institute of Particle Physics, Central China Normal University, Wuhan 430079, China
[3] School of Mathematics and Physics, China University of Geosciences, Wuhan 430074, China
[4] Nuclear Science Division MS 70R0319, Lawrence Berkeley National Laboratory, Berkeley, CA 94720, USA
* Correspondence: bwzhang@mail.ccnu.edu.cn

Abstract: Reconstructed jets initiated from heavy quarks provide a powerful tool to probe the properties of the quark–gluon plasma (QGP) and to explore the mass hierarchy of jet quenching. In this article, we review the recent theoretical progresses on heavy-flavour jets in high-energy nuclear collisions at the RHIC and LHC. We focus on the yields and substructures of charm and bottom quark jets with jet-quenching effects, such as the nuclear modification factors, transverse momentum imbalance, angular correlation, radial profiles, fragmentation functions, the "dead-cone" effect, etc.

Keywords: quark–gluon plasma; jet quenching; high-energy nuclear collisions; heavy-flavour jet

1. Introduction

High-energy nuclear collisions at the Relativistic Heavy Ion Collider (RHIC) and Large Hadron Collider (LHC) have opened up new avenues for the search for strongly interacting nuclear matter, the quark–gluon plasma (QGP) [1–4]. Investigating the formation of the QGP deepens our understanding of quantum chromodynamics (QCD) under extreme conditions at high temperature and density [5,6] and the evolution of the Universe at the first microsecond [7]. The jet-quenching phenomena, the energy attenuation of fast partons due to their strong interactions with the QCD medium, provide an army of powerful tools to study the properties of the QGP, such as the yield suppression of high-p_T hadron/jet, the p_T asymmetry of dijets, γ/Z^0+ jets as well as jet substructures [8–21].

In elementary proton–proton reactions, the productions of charm and bottom quarks are perturbatively calculable, since their large masses ($M_c \sim 1.5$ GeV, $M_b \sim 4.8$ GeV) act as a natural cut-off above the Λ_{QCD} [22]. Heavy quarks are produced in the initial hard scattering at a very early stage due to their large masses, therefore witnessing the whole QGP evolution. Meanwhile, while their thermal production is almost negligible with the initial conditions so far accessible in heavy-ion programs at the RHIC and LHC [23], the productions of charm and bottom hadron/jets make a very promising hard probe to the transport properties of hot and dense quark matter. During the past decade, the experimental measurements including the nuclear modification factor R_{AA} [24–30] and the collective flow (the direct flow v_1 [31,32] and elliptical flow v_2 [33–36]) of heavy-flavour hadrons both at the RHIC and LHC have attracted much attention from the community of high-energy nuclear physics.

A lot of theoretical studies have been performed to confront the experimental data obtained in the high-energy heavy-ion collisions, which greatly improve our understanding of the in-medium evolution [37–58] and hadronization mechanisms [59–61] of heavy quarks (for detailed reviews see [62–70]). Specifically, the current models treat the elastic and inelastic interactions between heavy quarks and the QGP medium with multiple methods, consisting mainly the perturbative or non-perturbative analytic calculations (SCET [41,71], CUJET [46,72], DREENA [73–78], WHDG [79,80], AdS/CFT (HG) [81,82]), and the Monte

Carlo transport approaches based on the Boltzmann (BAMPS [83–86], MC@$_s$HQ [87–89], (Q)LBT [47,57], LIDO [49,90], Catania-pQCD/QPM [91–94]), the Langevin (POWLANG [45,95,96], Duke [44,97], UrQMD [98–100], TAMU [40,101,102], SHELL [103–105]) and the Kadanoff–Baym (PHSD [106–108]) equations. These phenomenological studies reveal a fact that the elastic scattering of heavy quarks in the hot/dense nuclear matter is important, especially at the lower p_T region ($p_T^Q < 5m_Q$), different from our experience of treating light quarks or gluons. One of the central issues of investigating the heavy-flavour production in the heavy-ion program is extracting the diffusion coefficient D_s, which is directly related to the transport properties of the hot QCD matter. Additionally, different from the fragmentation hadronization of heavy quarks in a vacuum, within the hot and dense nuclear matter, the heavy-flavour hadrons can be produced by a combination of heavy quarks and thermal partons. Such a coalescence hadronization mechanism plays an important role in the collective flow [33,34] and baryon-to-meson ratio [109,110] of charmed hadron in nucleus–nucleus collisions at the RHIC and LHC.

In recent years, the experimental measurements on heavy-flavour jet (a reconstructed jet containing a heavy quark or a heavy-flavour hadron) have made great strides in p+p [111–118], p+A [111,119,120] and A+A collisions [121–128]. A wealth of information carried by heavy-flavour jets not only offers a new topic of jet physics and the application of the perturbative QCD, but also their medium modifications in heavy-ion collisions are also of great significance to reveal the in-medium energy loss mechanism of heavy quarks, to address the mass effect of jet quenching, and to extract the transport properties of the QGP.

2. Recent Advances of Heavy-Flavour Phenomenology in Heavy-Ion Collisions

Generally speaking, as we discussed in the last section, the reason for treating the heavy flavours as powerful hard probes to the transport properties of the QGP consists of at least three aspects. Firstly, the large mass ($M_Q \gg \Lambda_{QCD}$) makes it available to compute the differential cross-section of heavy quarks in the binary nucleon–nucleon collisions based on the perturbative QCD (pQCD) scheme within the next-to-next-to-leading order (NNLO) precision [129]. Secondly, due to the large mass ($M_Q \gg T_{\rm med}$), the total yield of heavy quarks in nucleus–nucleus collisions only depends on their initial production at hard scattering. Since the momentum transfer of the in-medium collisions $q^2 \sim g^2T^2$ ($T \sim$ 0.4–0.5 GeV) is much smaller than the creation energy of heavy quark pairs at the current collision energy, both at the RHIC and LHC, the subsequent contribution from the thermal creation during the QGP evolution is negligible [23]. Apart from this, according to the Heisenberg uncertainty principle, the formation time of heavy quarks ($\tau_0 \sim \frac{1}{2m_Q} < 0.1$ fm/c) is shorter than the formation time of the quark–gluon plasma ($\tau_f \sim 0.6$ fm/c). Therefore heavy quarks witness the entire evolution of the hot/dense nuclear matter until the freeze-out. In this section, we will briefly introduce the recent theoretical advances that help us understand the heavy-flavour production in heavy-ion collisions, including mainly the following several aspects, the initial production, the transport approaches, the hadronization mechanisms, and the extraction of diffusion coefficient.

2.1. Production of Heavy Quarks in p+p Collisions

The production of heavy quarks in proton–proton collisions establishes a baseline to investigate the nuclear modification in high-energy nuclear collisions both at the RHIC and LHC. The yield of heavy flavours in nucleus–nucleus collisions generally is viewed as the sum of that in $N_{\rm coll}$ binary nucleon–nucleon collisions while taking into account the initial cold nuclear matter effect (usually considered by using the nuclear-modified parton distribution function [130–132]). In the fixed-flavour-number scheme (FFNS) [22], the cross-section of heavy quarks in p+p collisions can be expressed based on the factorization theorem,

$$d\sigma_Q[s, p_T, y, m_Q] \simeq \sum_{i,j} \int_0^1 dx_i \int_0^1 dx_j f_i^A(x_i, \mu_F) f_j^A(x_j, \mu_F) d\tilde{\sigma}_{ij\to Q+X}[x_i, x_j, s, p_T, y, m_Q, \mu_F, \mu_R] \quad (1)$$

where s is the square of the centre-of-mass energy of the incoming proton, p_T is the transverse momentum of the produced heavy quark, and y is the rapidity. f_i^A (f_j^B) is the parton distribution function (PDF) quantifying the probability to find a parton with flavour $i(j)$ and carrying momentum fraction $x_{i(j)}$ in the colliding proton $A(B)$, which relies on the factorization scale μ_F. $\tilde{\sigma}_{ij\to Q+X}$ represents the cross-section of the partonic hard process $i + j \to Q + X$ that can be calculated relying on the pQCD. The partonic cross-section $\tilde{\sigma}_{ij\to Q+X}$ also relies on the strong coupling constant α_s determined at the renormalization scale μ_R. Note that Equation (1) sums all partonic hard processes $i + j \to Q + X$, where i, j are the active flavours including $(u, \bar{u}, d, \bar{d}, s, \bar{s}, g)$ but not heavy quarks. Only at the factorization scale $\mu_F > m_c$, can charm be viewed as an active flavour, often used for beauty production. The differential cross-section $d\sigma_Q$ can be convolved with a scale-independent fragmentation function $D_Q^H(z)$, such as the Peterson [133] or Lund [134] forms, to obtain the cross-section of the heavy-flavour hadron,

$$d\sigma_H = d\sigma_Q \otimes D_Q^H(z) \quad (2)$$

where H denotes the heavy-flavour hadron and z the momentum fraction carried by H.

Since the FFNS is usually applicable at the low p_T region ($0 < p_T < 5m_Q$), for the higher kinematic region ($p_T \gg m_Q$), the logarithmic terms ($\frac{\alpha_s}{2\pi} ln(p_T^2/m_Q^2)$) in the perturbative expansion of the cross-section become large, and should be resummed to all orders. To implement such a resummation, one has to absorb the large logarithmic terms into the parton distribution function and fragmentation function. This treatment requires that heavy quarks are active flavours when the factorization scale is $\mu_F > m_Q$. In other words, such a scheme has a variable number of active flavours when μ_F crosses the heavy quark mass, hence named the variable-flavour-number scheme (VFNS). In particular, when the heavy quark mass can be neglected in the evaluation of the short-distance cross-section, the VFNS scheme is called the zero-mass VFNS (ZM-VFNS). In the ZM-VFNS, the differential cross-section of a heavy-flavour hadron based on the factorization theorem can be expressed as:

$$d\sigma_{H+X} \simeq \sum_{i,j} \int_0^1 dx_i \int_0^1 dx_j f_i^A(x_i, \mu_F) f_j^A(x_j, \mu_F) d\tilde{\sigma}_{ij\to k+X} D_k^H(z, \mu'_F) \quad (3)$$

where $D_k^H(z, \mu'_F)$ is given by a convolution of a perturbative-fragmentation function (PFF) $D_k^Q(z, \mu'_F)$ describing a parton k fragmentation into heavy quark Q, with a scale-independent one $D_Q^H(z)$ for the hadronization of a heavy quark. Note that in Equation (3) the sum covers all possible partonic-hard processes ($i + j \to k + X$) where i, j, k can be light quarks, gluons, and heavy quarks [135]. Since heavy quark mass is neglected in the computation of the cross-section, the ZM-VFNS is expected to be reliable only at very high p_T.

To find a unified theoretical framework that combines the advantages of the FFNS at low p_T and the ZM-VFNS at high p_T, in recent years the interpolation schemes have been established, such as the general-mass VFNS (GM-VFNS) [136,137] and the fixed-order plus next-to-leading logarithms (FONLL) [129,138]. For instance, by using an interpolating function $G(m_Q, p_T) = p_T^2/(p_T^2 + c^2 m_Q^2)$ where c is set to $c = 5$, the FONLL scheme can well describe the heavy-flavour production in the entire kinematic region. For more details of the interpolation schemes see [22] and the references therein.

Compared to the analytic calculation schemes discussed above, the general-purpose Monte Carlo event generator, such as PYTHIA [139], HERWIG [140], POWHEG [141] and SHERPA [142], can provide a more complete description of all the final-state particles at the parton or hadron level. Especially for the studies of jet physics, the Monte Carlo event generators can give more precise descriptions of the observations relating to the jet substructure than analytic calculations.

2.2. Transport of Heavy Quarks in the QGP

Transport approaches are wildly used in the current theoretical studies of heavy-flavour production in high-energy nuclear collisions. At the lower p_T region, the elastic scattering of heavy quarks with the thermal parton (light quark or gluon) has been proven to be the dominant mechanism of energy loss. Generally, the kinetic theory based on the Boltzmann transport equation is a popular treatment for in-medium heavy quark evolution. The Boltzmann equation for the distribution function of heavy quarks can be written in a compact form,

$$p_\mu \partial_\mu f_Q(x,p) = C[f_q, f_{\bar{q}}, f_g, f_Q](x,p) \tag{4}$$

where $f_Q(x,p)$ is the phase-space distribution of heavy quarks. In the QGP, the phase-space distributions of light quark f_q and gluon f_g can be solved by the Boltzmann equation [143,144]. Subsequently, the relativistic Boltzmann-like collision integral $C[f_Q](x,p)$ has a simplified form [42,145],

$$C[f_Q] = \int d^3q[\omega(\mathbf{p}+\mathbf{q},\mathbf{q})f_Q(\mathbf{x},\mathbf{p}+\mathbf{q},t) - \omega(\mathbf{p},\mathbf{q})f_Q(\mathbf{x},\mathbf{p},t)] \tag{5}$$

where $\omega(\mathbf{p}+\mathbf{q},\mathbf{q})$ represents the transition rate of a heavy quark from the momentum $\mathbf{p}+\mathbf{q}$ to $\mathbf{p}$ by collisions with quasiparticles. This rate is usually determined by the matrix elements of the $2 \to 2$ QCD scattering. With the assumption that the momentum transfer $|\mathbf{q}|$ is small compared to the momentum of a heavy quark, we can expand $\omega(\mathbf{p}+\mathbf{q},\mathbf{q})f_Q(\mathbf{x},\mathbf{p}+\mathbf{q},t)$ around $\mathbf{q}$ by utilizing the Taylor formula to obtain the Fokker–Planck equation,

$$\frac{\partial f_Q}{\partial t} = \frac{\partial}{\partial p_i}\left[A_i(\mathbf{p})f_Q + \frac{\partial}{\partial p_j}[B_{ij}(\mathbf{p})f_Q]\right] \tag{6}$$

where two coefficients $A_i(\mathbf{p}) = \int d^3q\omega(\mathbf{p},\mathbf{q})q_i$ and $B_{ij}(\mathbf{p}) = \int d^3q\omega(\mathbf{p},\mathbf{q})q_iq_j$ are directly related to the drag coefficient (η_D) and the momentum diffusion coefficient (κ), which control rate of the energy loss and the momentum broadening of heavy quarks in the hot medium, respectively. Indeed, the Fokker–Planck equation is equivalent to another more well-known equation, the Langevin equation,

$$\frac{d\vec{x}}{dt} = \frac{\vec{p}}{E} \tag{7}$$

$$\frac{d\vec{p}}{dt} = -\eta_D(p)\vec{p} + \vec{\xi}(t) \tag{8}$$

where the stochastic term $\vec{\xi}(t)$ describes the random kicks suffered in heavy quarks from the medium constituents, which obeys a Gaussian distribution with a mean value 0 and variance κ. The drag coefficient η_D and the diffusion coefficient κ are related by the fluctuation–dissipation theorem (FDT) $\kappa = 2\eta_D ET$. Note that at higher kinematic regions ($p_T^Q > 5m_Q$), the medium-induced gluon radiation plays an increasingly important role in the energy loss of heavy quarks. The radiative energy loss of heavy quarks is treated with various formalisms and at different approximations [53,146–150], which usually provide the radiated gluon spectra as a function of momentum fraction x and transverse momentum $k_\perp$. In the Langevin equation, the radiative energy loss of heavy quarks can be coupled with the collisional one by adding a recoil term $-\vec{p}_g$ caused by the radiated gluon [44]. The four-momentum of the radiated gluon can be easily sampled based on the radiation spectra $dN_g/dxdk_\perp^2$.

In many of the recently developed theoretical frameworks modelling the production of heavy flavour in heavy-ion collisions, the Boltzmann and Langevin equations are the two most popular choices, especially for Monte Carlo simulations. Concerning the performance of these two approaches, detailed comparisons have been discussed in [151,152]. In general, the implementation of the Boltzmann equation implies that the medium consists of well-defined quasiparticles, while the Fokker–Planck (Langevin) equation is realized in a more general way without the quasiparticle assumption. However, the advantage of the Boltzmann equation is that it can naturally describe the heavy quark evolution even under off-equilibrium conditions, which may be the case of the early pre-equilibrium stage in heavy-ion collisions [153].

2.3. Hadronization: Fragmentation and Coalescence

Studying the yield suppression and collective flow of heavy-flavour hadrons also deepens our understanding of heavy quark hadronization in nucleus–nucleus collisions, which shows different mechanisms with that in a vacuum. As discussed in Section 2.1, fragmentation functions describe the non-perturbative hadronization process of heavy quarks into heavy-flavour hadron in a vacuum. The most commonly used fragmentation function is the Peterson form [133],

$$D_{H/Q}(z) = \frac{N}{z[1 - \frac{1}{z} - \frac{\epsilon_Q}{1-z}]} \tag{9}$$

where z denotes the momentum fraction carried by the heavy hadron from the heavy quark in the fragmentation process ($0 < z < 1$), which implies that the heavy hadron must have smaller energy than the heavy quark. The only tunable parameter in Equation (9) is ϵ_Q that can be determined by fitting to the measured spectra of the heavy-flavour hadrons. N is the normalization factor to guarantee $\int_0^1 dz D_{H/Q}(z) = 1$.

Measurements on the collective flow [33,34] and baryon-to-meson ratio [109,110] of charmed hadron A+A collisions suggest the existence of a new hadronization mechanism, coalescent of heavy quarks. The basic idea behind the coalescence mechanism is that a heavy quark can combine with a light anti-quark from the medium when they have enough small distance in the coordinate-momentum space. It means that the heavy-flavour meson has larger energy than the parent heavy quark, differing from the mechanism of fragmentation. The distribution function of the formed heavy-flavour meson usually can be obtained by a convolution with the following schematic form.

$$f_M \sim g_M f_{Q(\bar{Q})} \otimes f_{\bar{q}(q)} \otimes \phi_M \tag{10}$$

where g_M denotes the degeneracy of the heavy-flavour meson in spin and isospin, $f_{Q(\bar{Q})}$ and $f_{\bar{q}(q)}$ are the distribution functions of the heavy and light quarks in the coordinate-momentum space, respectively. ϕ_M represents the Wigner transform of the wave function of the heavy-flavour meson, commonly approximated by the ground state one of the simple harmonic oscillators [47].

In the realistic implementation of the heavy quark hadronization in nuclear collisions, the first step is to determine the probability of coalescence by integrating the distribution function of Equation (10). If coalescence occurs, one can sample a light anti-quark based on the thermal equilibrium distribution, otherwise Equation (9) is used to fragment the heavy quark into a hadron. At least in the lower p_T region, the experimental results favour the coalescence mechanism [94]. The coalescence of heavy quarks seems to decrease the suppression factor and enhance the collective flow of heavy-flavour hadrons, especially at $p_T < 6$ GeV. The recent studies [59,61] show that the coalescence mechanism is important in the description of the Λ_c/D^0 ratio measured by the STAR [109] and ALICE [110] collaborations. Additionally, the hadronic scattering between the D meson and light-flavour hadrons ($D - \pi$, $D - \rho$) has also been studied in [154], but its influence on the D meson R_{AA} was found to be very limited [155].

2.4. Extraction of the Diffusion Coefficient of Heavy Quarks

One of the most important goals of the heavy-ion collision experiment is to investigate the transport properties of the QCD matter under extremely hot and dense conditions. As discussed above, due to the large mass ($m_Q \gg T_{\rm med}$), heavy quarks are believed to be powerful tools for exploring the transport properties of the QGP. Phenomenological studies of heavy-flavour production in high-energy nuclear collisions provide a unique opportunity to extract the transport coefficient of the QGP, such as the momentum diffusion coefficient κ of heavy quarks, whose longitudinal and transverse components can be convenient to define as,

$$\kappa_{||} \equiv -\frac{d\left\langle(\Delta p_{||})^2\right\rangle}{dt} \tag{11}$$

$$\kappa_{\perp} \equiv \frac{1}{2}\frac{d\langle(\Delta p_{\perp})^2\rangle}{dt} \tag{12}$$

where $\Delta p_{||}$ and $\Delta p_{\perp}$ momentum changes parallel and perpendicular to the heavy quark formulation. By definition, $\kappa_{\perp}$ can be directly related to the jet transport coefficient $\hat{q}$ which quantifies all the transverse momentum broadening of hard partons as traversing the QGP medium. Assuming that the κ is isotropic, namely, $\kappa_{\perp} = \kappa_{||} = \kappa$, one can obtain a simplified relation $\hat{q} = 2\kappa$. This relation has been employed in the modified Langevin equation to balance the two parts of the contribution from the collisional and radiative energy loss of heavy quarks [44,156]. Here we only overview the recent advances of the κ extraction by different model calculations. A more detailed and profound discussion about this topic can be found in [64].

The momentum diffusion coefficient κ can be easily converted to the spatial one D_s with the relation $\kappa = 2T^2/D_s$. In recent years, the temperature dependence of the dimensionless quantity $2\pi TD_s$ has been estimated by a lot of theoretical frameworks, such as the lattice QCD (lQCD) [157–159], LO pQCD [43,160], QPM calculations [92], T-matrix [40], PHSD [161], MC@$_s$HQ [22], Ads/CFT [162], duke (Bayesian analysis) [163], and hadronic matter [102,164], as shown in Figure 1.

The estimates by the lQCD from the first principles provide a valuable reference for the model extractions of $2\pi TD_s$. As one can see, with relatively large uncertainties, the lQCD calculations in the quenched approximation give $D_s \sim 3.7$–7.0 [159] over the temperature range from T_{pc} to $2T_{pc}$. However, it is difficult to extract meaningful information about the temperature dependence of $2\pi TD_s$ from the current lQCD results. Furthermore, except for the pQCD calculations at the leading-order which show obvious larger values than others, these extractions of $2\pi TD_s$ based on the recently developed models are consistent with the lQCD data, as well as previous studies presented in [64] which give $2\pi TD_s \sim 2$–4 near the critical temperature. Although these calculations give different values of $2\pi TD_s$ versus T/T_{pc}, most estimations show that D_s slightly increases with T. It implies that the interactions between a charm quark and the QCD medium have the strongest strength near the critical temperature. However, no direct evidence has been found in the experiment to verify this upward trend of D_s so far because it is hard to find an observation only sensitive to the in-medium interactions at the late stage of the QGP evolution. Fortunately, the data-driven analysis utilizing Bayesian inferences seems to shed new light on this issue. The temperature and momentum dependence of D_s has been extracted from the available experimental data (R_{AA} and v_2 of a D meson both at the RHIC and LHC) [163] based on the Duke–Langevin transport model, which indeed shows an upward trend of $2\pi TD_s$. More recently, this approach of Bayesian inference has been improved with the help of information field theory [165,166] in [167]. Therefore, one can now extract model parameters without relying on an explicit form of parametrization, leading to a robust determination by such a model–data fit.

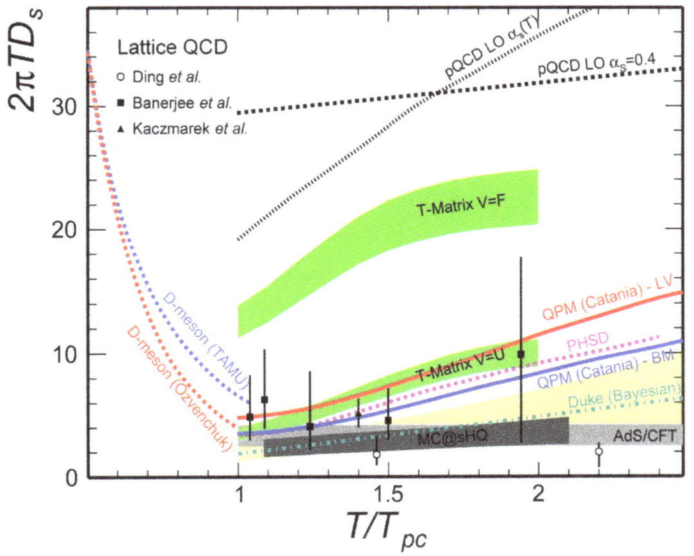

Figure 1. Spatial diffusion coefficient ($2\pi T D_s$) of charm quark in the quark–gluon plasma calculated by different approaches versus the reduced temperature (T/T_{pc}). The lattice QCD calculations in the quenched approximation [157–159] are compared with the estimations based on different models [22,40,43,92,102,160–164]. The figure is from [66].

3. Production of Heavy-Flavour Jets in Heavy-Ion Collisions

3.1. Nuclear Modification Factors of Production Yields

To address the nuclear effect in relativistic heavy-ion collisions, the nuclear modification factor R_{AA} is conventionally utilized to quantify the yield suppression of hadron/jet in A+A collisions per binary nucleon–nucleon collision relative to p+p [168],

$$R_{AA} = \frac{1}{\langle N_{\rm bin}^{\rm AA}\rangle} \frac{d\sigma^{\rm AA}/{\rm dydp_T}}{d\sigma^{\rm pp}/{\rm dydp_T}} \tag{13}$$

where the scaling factor $\langle N_{\rm bin}^{\rm AA}\rangle$ denotes the number of binary nucleon–nucleon collisions in A+A [169]. It has been observed that the values of R_{AA} of hadrons and jets are smaller than one in nucleus–nucleus collisions both at the RHIC [170–172] and LHC [173,174], and these measurements could be explained by the mechanism of partonic energy loss, which in turns serve as convincing evidence for the formation of QGP in such extremely hot and dense conditions. Meanwhile, the jet transport parameter $\hat{q} \equiv d\langle p_\perp^2\rangle/dL$ [175] representing the strength of in-medium partonic interactions could be extracted from the available R_{AA} data by various theoretical models [167,176–179].

Additionally, to test the mass dependence of jet quenching, the R_{AA} has also been used in a comparison of the yield suppression between heavy-flavour jets and inclusive jets. Benefiting from the fact that heavy-flavour jets are produced abundantly as the centre-mass energy increases in hadronic collisions at the LHC, the exploration of a heavy quark-tagged jet produced in heavy-ion collisions has gradually attracted much attention. The first experimental effort focused on the production of a b-jet was implemented by the CMS collaboration [121] in 2013, as shown in the left plot of Figure 2, where a b-jet is defined as jets containing at least one B hadron inside the jet-cone. The red points are the CMS data and the coloured bands are the theoretical calculations. This measurement accounts for the b-jet samples in minimum bias collisions (0–100%). We note that even with large experimental uncertainties, the b-jet R_{AA} slightly increases with jet p_T and varies from 0.4 to

0.8. Significant suppression of the b-jet yield in Pb+Pb collisions at $\sqrt{s_{NN}}$ = 2.76 TeV relative to the p+p baseline was observed for the first time, which indicates that bottom quarks strongly interact with the hot/dense nuclear matter. Furthermore, within experiment uncertainties, the results were found to be consistent with the pQCD-based calculations conducted in [180] when the coupling factor g^{med} varied from 1.8 to 2.2.

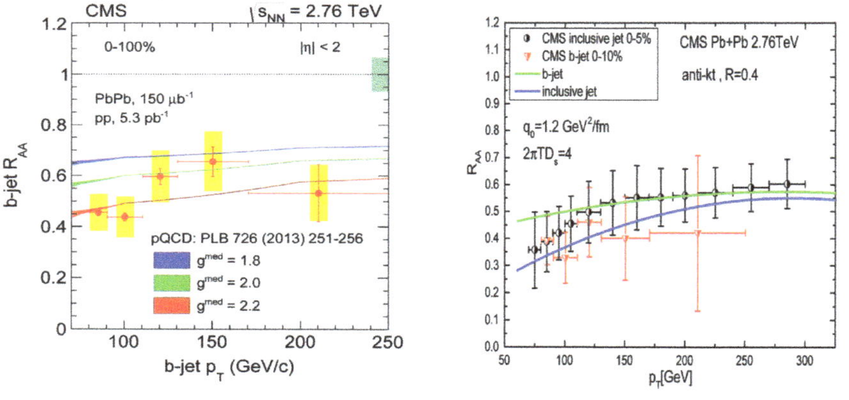

Figure 2. Left: the measured nuclear modification factor R_{AA} of the inclusive b-jet versus b-jet p_T by the CMS collaboration in Pb+Pb at $\sqrt{s_{NN}}$ = 2.76 TeV for minimum bias collisions [121]. **Right**: the comparison of R_{AA} between the inclusive jets and b-jets versus jet p_T in central Pb+Pb collisions at $\sqrt{s_{NN}}$ = 2.76 TeV [104]. The figures are from [104,121].

To address the difference of the yield suppression between the b-jet and inclusive jet (mainly initiated by a massless light quark or gluon), a direct comparison of their R_{AA} in the right plot of Figure 2 was presented by the SHELL approach, which applies a Langevin transport model to describe heavy quark propagation in the QGP [104] in central Pb+Pb collisions at $\sqrt{s_{NN}}$ = 2.76 TeV, as well as the next-to-leading order pQCD calculations matched with the parton shower effect for the p+p baseline [142,181]. In the model, the jet transport parameter $\hat{q}$ was extracted by the production of an identified hadron in A+A collisions [182], and then the spatial diffusion coefficient D_s of heavy quarks can be determined by the D meson R_{AA} data [26,127]. The measured R_{AA} of the inclusive jet with the centrality of 0–5% [183] and b-jet with 0–10% [121] are also illustrated in the plot of Figure 2. Although the R_{AA} of the b-jet seems to be slightly smaller than that of the inclusive jet, the CMS collaboration claims that no clear difference of R_{AA} between the inclusive jet and b-jet was found, because the current uncertainties of the b-jet data are too large. However, the theoretical calculations in [104] suggest that b-jet R_{AA} may be larger than inclusive jet R_{AA}, due to the "dead-cone" effect of the bottom quarks, which suppresses the medium-induced gluon radiation of massive heavy quarks within a cone $\theta \sim M/E$ [184]. A more precise measurement is necessary to resolve the tension between the experimental data and theoretical calculations. It is very exciting the fact that recently, the ATLAS collaboration reported preliminary results by simultaneously measuring the R_{AA} of the inclusive jet and b-jet in 0–20% Pb+Pb collisions at $\sqrt{s_{NN}}$ = 5.02 TeV [124], which shows a clear weaker suppression of the b-jet, and the features can be described by the theoretical calculations [49,104]. Although the mass hierarchy of jet quenching at the particle level has been confirmed by a lot of experimental data [185,186], it is indisputable that the ATLAS measurement makes a crucial step towards finding the mass effect at the jet level. The comparison of the c- and b-jet R_{AA} has been presented in [71] with the SCET model [187,188], which shows no significant difference at $p_T > 50$ GeV. More recently, some exploratory estimates indicate that the R_{AA} of the c-jet may be stronger than that of the inclusive jet at higher jet p_T due to their different constituents [189,190], an interesting finding to be investigated further in detail.

Beyond R_{AA}, another observation I_{AA} [16] has also been utilized to study the yield suppression of b-jets tagged by Z^0 bosons in high-energy nuclear collisions [191]. Similar to R_{AA}, I_{AA} is defined as follows,

$$I_{AA} = \frac{1}{\langle N_{\rm bin}^{\rm AA} \rangle} \frac{\frac{d\sigma^{\rm AA}}{dp_{\rm T}^{\rm jet}}|_{p_{\rm T}^{\rm min}<p_{\rm T}^{Z}<p_{\rm T}^{\rm max}}}{\frac{d\sigma^{\rm pp}}{dp_{\rm T}^{\rm jet}}|_{p_{\rm T}^{\rm min}<p_{\rm T}^{Z}<p_{\rm T}^{\rm max}}} \tag{14}$$

One can see that I_{AA} quantifies the yield variation of the jet in A+A collisions per binary nucleon–nucleon collision relative to the p+p baseline, after integrating the Z^0 boson p_T. It has been proposed that the associated production Z^0 + b-jet may be helpful in addressing the mass dependence of the jet-quenching effect, since the Z^0 + jet processes significantly exclude the contamination of gluon-initiated jets [192]. Thereby the comparison of I_{AA} between the Z^0 + jet and Z^0 + b-jet can provide direct features of the mass effect of heavy quark jets compared to light-quark jets. Figure 3 shows the comparisons of the calculated I_{AA} of the Z^0 + b-jet and Z^0 + jet in central 0–10% Pb+Pb collisions at $\sqrt{s_{NN}}$ = 5.02 TeV, where the jets are reconstructed by the anti-k_T algorithm with a cone size R = 0.3 and pseudorapidity $|\eta^{\rm jet}| < 1.6$. The calculations are presented within different p_T windows of Z^0 bosons in the three panels, namely 40–60, 60–80, and 80–100 GeV. We note that the shapes of I_{AA} are flat at the panel of $40 < p_T^Z < 60$ GeV but have downward trends at that of $80 < p_T^Z < 100$, and the I_{AA} in the right panel even shows enhancement at $p_T^{\rm jet}$. This is because if one constrains the Z^0 meson momentum in the event selection, such as $80 < p_T^Z < 120$ GeV, then the selected jet distribution with p_T would fall steeper at $p_T^{\rm jet} < 80$ GeV, which naturally leads to a relatively large value of the nuclear modification factors at $p_T^{\rm jet} < 80$ GeV, even larger than one. Additionally, one observes that at each panel the model calculations show that I_{AA} of the Z^0 + b-jet is visibly higher than that of the Z^0 + jet in nucleus–nucleus collisions, which indicates that the Z^0-tagged light-quark jets lose more energy than the Z^0-tagged b-jets traversing the QGP. These comparisons would be helpful to directly test the mass effect of jet energy loss in heavy-ion collisions at the LHC from a new perspective.

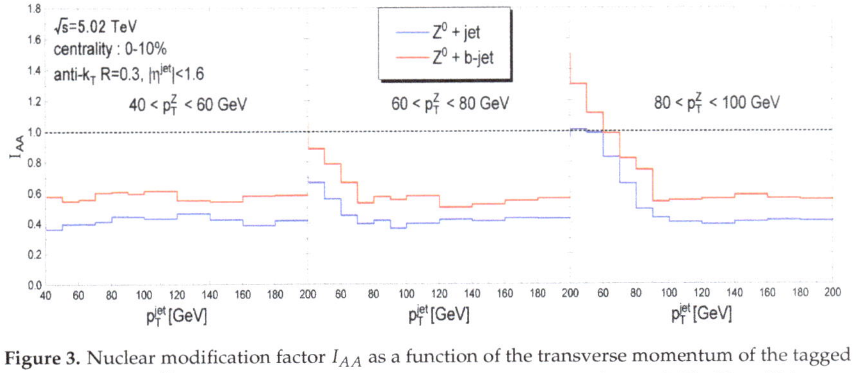

Figure 3. Nuclear modification factor I_{AA} as a function of the transverse momentum of the tagged jet within three p_T^Z windows: 40–60, 60–80, 80–120 GeV in central 0–10% Pb+Pb collisions at $\sqrt{s_{NN}}$ = 5.02 TeV [191].

3.2. Transverse Momentum Imbalance

The transverse momentum imbalance ($x_J = p_{T,2}/p_{T,1}$) is another useful observation, describing the momentum asymmetry of the dijet system in the transverse plane, where $p_{T,1}$ and $p_{T,2}$ denote the leading and sub-leading jet p_T. It is noted that in the fixed-leading-order QCD calculations the two outgoing hard partons should be strictly back-to-back in the transverse (x_J = 1), but the higher-order corrections and vacuum shower may break the symmetry which leads to $x_J < 1$. In heavy-ion collisions, the smaller x_J of the γ+jet [193] and Z^0+jet [194] systems have been observed in Pb+Pb collisions compared to p+p, which results from the energy loss of the tagged jet. The CMS collaboration reports the measurement on x_J of the inclusive and $b\bar{b}$ dijets in Pb+Pb collisions at $\sqrt{s_{NN}}$ = 5.02 TeV [122]. In their measurements, the biggest challenge was how to select the $b\bar{b}$ dijet events initiated by the hard heavy-quark pairs, because it is crucial to address the mass effect by directly comparing such heavy-quark dijets with inclusive dijets. On the theoretical side, the production mechanisms of heavy quarks can be categorized into three classes: flavour creation (FCR), flavour excitation (FEX), and gluon splitting (GSP) [195–198], only FCR represents the dijets initiated by heavy-quark pairs originating from the hard process. The CMS collaboration suggests a strategy to separate the FCR processes by selecting $b\bar{b}$ dijets that have a large opening angle ($|\Delta\phi| > 2\pi/3$) in azimuth, which could significantly suppress the contributions of the other two. This method has also been used in theoretical studies [104,199,200]. Figure 4 shows a comparison of the averaged x_J of the inclusive and $b\bar{b}$ dijets in both p+p and Pb+Pb collisions with different centrality bins at $\sqrt{s_{NN}}$ = 5.02 TeV, as well as the experimental data [122], where $\langle x_J \rangle$ was estimated as follows.

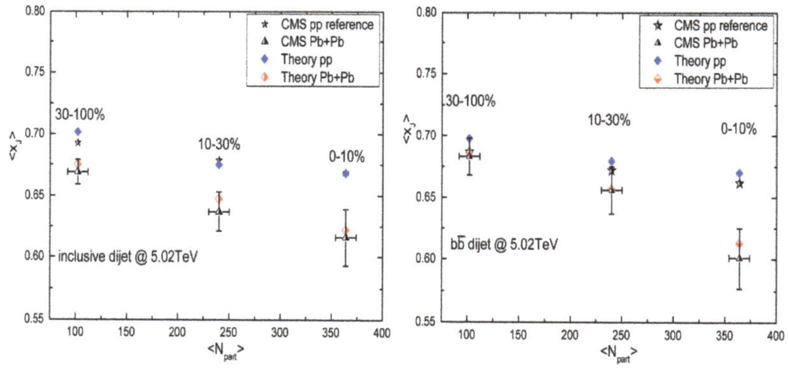

Figure 4. Averaged x_J value as a function of the number of participants calculated in p+p and Pb+Pb collisions $\sqrt{s_{NN}}$ = 5.02 TeV within different centrality bins compared with the experimental data, both for inclusive (**left**) and $b\bar{b}$ (**right**) dijets. Figures are from [104].

$$\langle x_J \rangle = \frac{1}{\sigma}\int_0^1 \frac{d\sigma}{dx_J} dx_J \tag{15}$$

The black triangle points are the CMS data in Pb+Pb collisions, and the black star points are the p+p reference used in their measurements. The blue and red rhombus points are the theoretical calculations, while the p+p reference is provided by the Monte Carlo event generator SHERPA [142] which matches the next-to-leading order QCD matrix elements and the parton shower effect in a vacuum [201,202]. The $\langle x_J \rangle$ points of the inclusive (left panel) and $b\bar{b}$ (right panel) dijets are listed within three centrality bins which correspond to the different numbers of the participant in Pb+Pb collisions. In Figure 4, the theoretical calculations based on the SHELL model [104] show an overall decrease in $\langle x_J \rangle$ in Pb+Pb collisions relative to the p+p baseline both for the inclusive and $b\bar{b}$ dijets, consistent with the CMS data and indicates that the asymmetry between these two leading jets is amplified in A+A collisions. The reduction in $\langle x_J \rangle$ is centrality-dependent since the in-medium interaction is sensitive to the temperature and size of the QGP. It is even more important that the calculations show that the decrease in $\langle x_J \rangle$ of the $b\bar{b}$ dijets is slightly smaller than that of the inclusive dijets within the same centrality bins. These results suggest that dijets initiated by bottom quarks may suffer smaller energy loss compared to those initiated by light quarks or gluons. Furthermore, another study on the $b\bar{b}$ dijet in heavy-ion collisions [199] proposed that the invariant mass m_{jj} of the dijet system could be a novel observation sensitive to mass effects of jet quenching.

In addition to the dijet system, the transverse momentum imbalances of the Z^0 + jet ($x_{jZ} = p_T^{\text{jet}}/p_T^Z$) and Z^0 + b-jet ($x_{bZ} = p_T^{\text{b-jet}}/p_T^Z$) have also been investigated [191]. It was found that the Z^0-tagging requirement considerably decreased the contribution of gluon-jets by 40% in Z^0 + jets compared to the dijet sample, especially at a lower jet p_T. The comparison of the medium modification on the x_J of Z^0 + jet and Z^0 + b-jet may be suitable to address the mass effect of jet quenching. Figure 5 shows the distributions of the x_{jZ} (left) and x_{bZ} (right) both in p+p and 0–10% Pb+Pb collisions at $\sqrt{s_{NN}}$ = 5.02 TeV. In the calculations, the selected Z^0 bosons are required to have $p_T^Z > 60$ GeV. The tagged jets (b-jets) are reconstructed with the anti-k_T algorithm with a cone-size R = 0.3 and pseudorapidity $|\eta^{\text{jet}}| < 1.6$, required to have $p_T^{\text{jet}} > 30$ GeV. In particular, to guarantee that the Z^0 bosons and the tagged jets are back-to-back in the transverse plane, the Z^0 + jet or Z^0 + b-jet pairs are required to have a large opening angle in azimuth, $\Delta\phi_{jZ}(\Delta\phi_{bZ}) > 7\pi/8$. The differences of x_{jZ} (x_{bZ}) distributions in p+p and Pb+Pb collisions are also shown in the lower panels. Due to the jet energy loss, the x_{jZ} and x_{bZ} distributions shift towards smaller x_J values in Pb+Pb collisions relative to p+p. Furthermore, one can find in the lower panel that the variations of x_{bZ} are slightly smaller than that of x_{jZ}. More intuitive comparisons between the averaged x_{jZ} and x_{bZ} are listed in Table 1. Within the statistical errors, the results show that $\Delta\langle x_{jz} \rangle \sim 0.136$ is considerably larger than $\Delta\langle x_{bz} \rangle \sim 0.092$, consistent with the expectation that bottom jets lose less energy than light-quark jets.

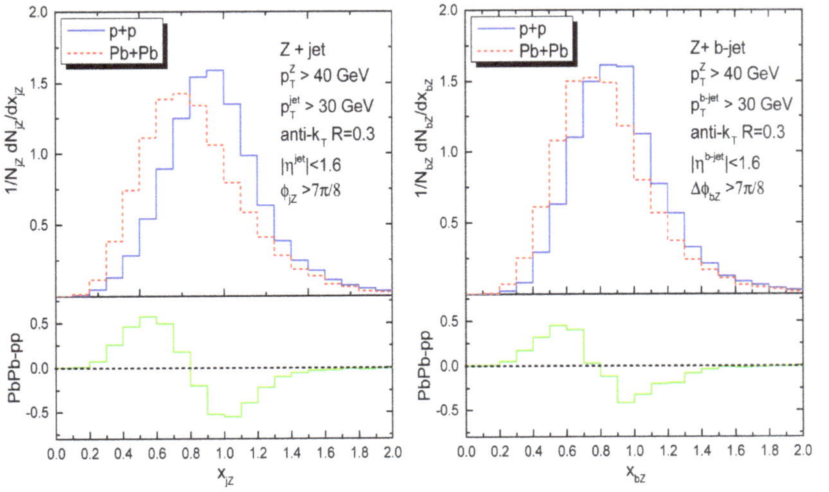

Figure 5. Distributions of x_{jZ} (x_{bZ}) of Z^0 + jet (**left**) and Z^0 + b-jet (**right**) both in p+p and 0–10% Pb+Pb collisions at $\sqrt{s_{NN}}$ = 5.02 TeV. The differences of x_{jZ} (x_{bZ}) distributions in p+p and Pb+Pb collisions (green line) are also shown in the lower panels. Figures are from Ref. [191].

Table 1. The averaged x_J of Z^0 + jet and Z^0 + b-jet both in p+p and Pb+Pb collisions at $\sqrt{s_{NN}}$ = 5.02 TeV, as well as their variations $\Delta x_J = \langle x_J \rangle_{pp} - \langle x_J \rangle_{PbPb}$. The statistical errors of x_J in the simulations are also presented. Table is from Ref. [191].

	Z^0 + jet	Z^0 + b-jet
$\langle x_J \rangle_{pp}$	0.987 ± 0.0047	0.941 ± 0.0056
$\langle x_J \rangle_{PbPb}$	0.851 ± 0.0061	0.849 ± 0.0064
$\Delta\langle x_J \rangle$	0.136 ± 0.0108	0.092 ± 0.012

3.3. Angular Correlation

Jet angular correlations, such as $\Delta\phi$ distribution of dijets [203,204] and γ/Z^0 + jet [205,206], are useful observable to address the medium-induced transverse momentum effect. In this context, estimating the medium modification on the angular distribution of heavy quark dijets in nucleus–nucleus collisions may also be of interest from the theoretical point of view. As shown in the left plot of Figure 6, medium modification of the azimuthal angular correlations ($\Delta\phi = |\phi_{b1} - \phi_{b2}|$) of the $b\bar{b}$ dijet system in Pb+Pb collisions with different centralities at $\sqrt{s_{NN}}$ = 5.02 TeV are calculated [207]. One can observe suppression at $\Delta\phi$ ∼0 and enhancement at $\Delta\phi \sim \pi$ in Pb+Pb collisions compared to the p+p, and the modifications are centrality dependent. Since the distributions are self-normalized, it implies that $b\bar{b}$ dijets with a larger opening angle (back-to-back) suffer relatively weaker yield suppression compared to that with a smaller one (collinear). It can be noted that the main contribution of $b\bar{b}$ dijet production at smaller $\Delta\phi$ is from the GSP process while larger $\Delta\phi_{bb}$ from the FCR process. The two b-jets from the former process share the energy of the gluon and then usually have lower p_T than that from the latter process. As a result, the yield at the smaller $\Delta\phi$ region is more sensitive to the selection cut $p_T^{\mathrm{jet}} > 20$ GeV. Actually, in another study on the angular correlations of Z^0 + b-jet [191], it's found that initial average b-jet p_T distribution versus $\Delta\phi$ play a critical role, as shown in the right plot of Figure 6. We see that the ratio of PbPb/pp in the middle panel is flat, and the average b-jet p_T distribution is also flat. It's reasonable to guess that in Pb+Pb the azimuthal angle between b-jet and Z^0 has not been modified compared to p+p, and the overall suppression occurs at whole $\Delta\phi_{bZ}$ region. Of course, we can imagine that it is more difficult for high-p_T (>30 GeV) jets to be significantly deflected by the scattering with thermal parton.

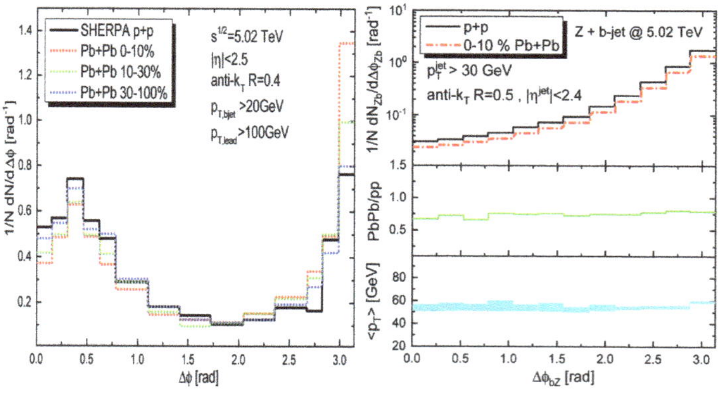

Figure 6. Left: normalized azimuthal angular distributions of $b\bar{b}$ dijet system in p+p and Pb+Pb collisions at $\sqrt{s_{NN}}$ = 5.02 TeV. Results for different centrality bins, 0–10%, 10–30%, 30–100%, are presented. **Right**: the azimuthal angular distribution of Z^0 + b-jet in p+p and 0–10% Pb+Pb collisions at $\sqrt{s_{NN}}$ = 5.02 TeV in the upper panel, while the ratio of PbPb/pp (green solid line) was shown in the middle panel and the averaged b-jet p_T (blue band) in the lower panel. Figures are from Refs. [191,207].

To probe the angular deflection caused by the in-medium p_T-broadening, observables accessible to lower p_T region are needed. For this reason, it's proposed in Ref. [105] that the heavy-flavour meson tagged by direct photon (γ+HF) may provide a promising channel, with several advantages: (1) the transverse momentum resolution of D^0 meson can be low down to $\sim$1 GeV [127] where the angular deflection is significant, (2) the photon gauges the initial momentum of heavy quarks, therefore, it's easy to quantify the direction change, (3) the selection bias effect can be suppressed by constraining the photon energy to citeCunqueiro:2021wls. In this way, the considerable angular de-correlations between the heavy quarks and photons are predicted both in central Au+Au collisions at the RHIC and Pb+Pb collisions at the LHC. Furthermore, by constructing the 2-dimensional $(\Delta\phi, x_J)$ correlation diagram of γ+HF, it's argued that the two aspects of jet quenching, energy loss, and p_T-broadening, can be well displayed simultaneously. Additionally, it's noted that another measurement on the angular correlations of D^0+hadron in Au+Au collisions at $\sqrt{s_{NN}}$ = 200 GeV may reflect the medium modification of the charm+jet correlation in the $\eta - \phi$ plane [208], that awaits further detailed investigations.

3.4. Radial Profile

The radial profile of the heavy-flavour jet represents the distribution of the angular distance $r = \sqrt{(\phi_Q - \phi_{jet})^2 + (\eta_Q - \eta_{jet})^2}$ between the heavy-flavour meson and the jet-axis in the $\eta - \phi$ plane. Systematic studies with a focus on the radial profiles of D-jet and B-jet in heavy-ion collisions are performed in Refs. [103,209]. As shown in the left panel of Figure 7, the model calculated radial profiles of D-jets both in p+p and 0–100% Pb+Pb collisions at $\sqrt{s_{NN}} = 5.02$ TeV compared to the CMS measurements [123]. The black and red triangle points represent the measured data. The D-jets are reconstructed with anti-k_T algorithm with R = 0.3 and $|\eta^{jet}| < 1.6$. All selected D-jets must satisfy $p_T^{jet} > 60$ GeV and contain at least one D^0 meson in jet-cone with $4 < p_T^D < 20$ GeV. The blue solid line is the p+p baseline provided by SHERPA [142], and the red dashed line denotes the calculations based on the SHELL model. One can observe that the model calculations show the radial profile of D-jets in Pb+Pb collisions shifts towards larger radii relative to that of p+p, which is consistent with the diffusion trend observed by the CMS collaboration. These results show a clear physics picture, that charm quarks change their moving direction when scattering with the thermal partons in the hot and dense QCD matter. The studies argue that the diffusion behavior of D meson is closely related to the p_T-broadening when charm quarks scatter with the thermal partons in the medium. It should be noted that in such an estimate the jets are required to have $p_T > 60$ GeV while D meson $p_T < 20$ GeV, which makes that the higher p_T jets can be viewed as a reference to probe the moving direction changes of charm quarks. It is found that the angular deviation $\Delta r = \sqrt{(\phi_c^f - \phi_c^i)^2 + (\eta_c^f - \eta_c^i)^2}$ of charm quarks from their initial position in the $\eta - \phi$ plane is p_T dependent, as shown in the right plot of Figure 7. The charm quarks with lower p_T are more likely to change their traveling direction via the in-medium scattering, and this feature also explains why no visible modification is observed in the CMS data for $p_T^D > 20$ GeV [123]. The angular deviation at lower p_T (<5 GeV) is dominated by elastic scattering, whereas at higher p_T by inelastic reactions. These investigations may cast light on the in-medium energy loss mechanisms and constrain the transport coefficients of heavy quarks from a new perspective. We notice that a preliminary result of the D-jet radial profile in Au+Au collisions at $\sqrt{s_{NN}} = 200$ GeV has been reported by the STAR collaboration in Ref. [125]. This result shows a similar diffusion effect of charm quark in jets in mid-central 10–40% collisions.

To test the mass effect reflected in the radial profile, an additional comparison of the medium modification between D-jet and B-jet has been presented in Refs. [210,211], where an inverse modification pattern on the radial profile of B-jets compared to D-jets is observed. The jet quenching effect seems to narrow the jet radial profiles of B-jets while broadening those of D-jets. It's demonstrated that the selection bias effect [212] in A+A collisions may play a pivotal role. Heavy quark jets with higher p_T have narrower initial radial distributions, and would naturally lead to narrower modifications when they fall into the lower p_T domain due to jet energy loss. This reveals the fact that the final-state modification of the jet is not only influenced by the pure medium effect, but also by the other factors, such as the initial spectra and the selection bias [5].

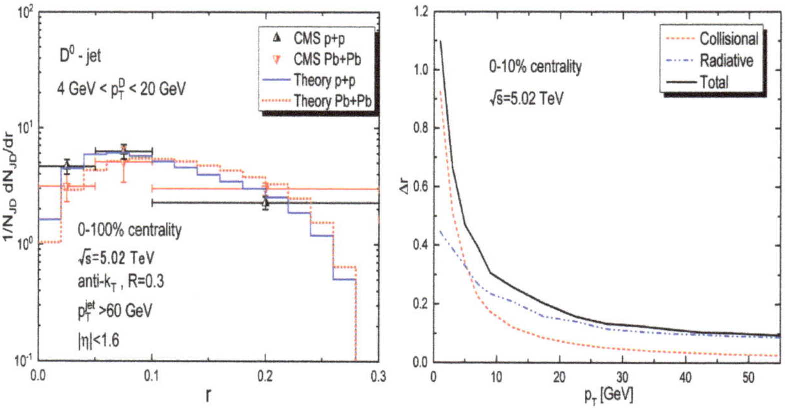

Figure 7. Left: radial profile of D-jet in p+p and Pb+Pb collisions at $\sqrt{s_{NN}}$ = 5.02 TeV. **Right**: angular deviation of charm quark as a function of initial p_T. Figures are from Refs. [103,209].

3.5. Fragmentation Function

The jet fragmentation function $D(z) = (1/N_{\text{jet}})dN_{\text{ch}}(z)/dz$ is one of the most well-explored jet substructure observable [213–215], which usually refers to the longitudinal momentum distribution of charged hadrons inside the jet-cone [216–219]. For heavy-flavour jets, the corresponding observable is the longitudinal momentum distribution of heavy-flavour mesons in jets, defined as in [112].

$$D(z_{||}) = \frac{1}{N_{\text{jet}}}\frac{dN_{\text{HQ}}(z_{||})}{dz_{||}}, \quad \text{where } z_{||} = \frac{\vec{p}_{\text{HQ}} \cdot \vec{p}_{\text{jet}}}{\vec{p}_{\text{jet}} \cdot \vec{p}_{\text{jet}}}\,. \tag{16}$$

On the one hand, the $D(z_{||})$ distribution may provide useful information to reveal the production mechanisms and substructure of heavy quark jets [220]. On the other hand, since $z_{||}$ denotes the momentum projection of the heavy-flavour hadron on the jet axis, the medium modification of the $D(z_{||})$ distribution in nucleus–nucleus collisions is closely related to the interplay of the partonic energy loss between the massive heavy quarks and the massless light partons [221]. Figure 8 shows the first theoretical investigation of the medium modification on the $D(z_{||})$ distributions of both D-jets and B-jets in Pb+Pb collisions at $\sqrt{s_{NN}}$ = 5.02 TeV. In these calculations, two jet p_T ranges are chosen, $5 < p_T^{\text{jet}} < 15$ GeV and $15 < p_T^{\text{jet}} < 50$ GeV. Respectively the selected D^0 (B^0) mesons are also required to have $p_{T,D^0(B^0)} > 2$ GeV and $p_{T,D^0(B^0)} > 5$ GeV. The black solid lines represent the p+p baseline of $D(z_{||})$ distributions calculated by the POWHEG+PYTHIA8 event generator [141,222–224], and the orange dash lines are the theoretical calculations in Pb+Pb collisions based on the SHELL model. The upper and middle panels correspond to the $D(z_{||})$ distributions of D-jets and B-jets, while the lower panels are their nuclear modification $D(z_{||})_{PbPb}/D(z_{||})_{pp}$ (green is D-jet and yellow B-jet). One can observe that the initial $D(z_{||})$ distributions in p+p are sensitive to the kinematic region of jet and heavy-flavour hadron, especially for D-jets. Moreover, even within the same kinematic region, a B-jet has an evident harder fragmentation pattern compared to a D-jet. The difference could be relevant to the fact that the stronger "dead-cone" effect suffered in heavier bottom quarks, in other words, the bottom quarks radiate less gluon and carry more energy fraction of jets than charm quarks. Besides, the contribution of the GSP process may also play different roles in the production of B-jets and D-jets, which may lead to additional differences in their $z_{||}$ distributions [221]. In nuclear collisions, the main finding is that the jet quenching effect results in softer fragmentation patterns of heavy-flavour jets in the QGP compared to that in a vacuum. It's different from what one could naively argue, that is, the energy fraction of heavy quarks in jets may increase because heavy quarks lose less energy than light partons. The modification of $D(z_{||})$ reveals the different energy loss mechanisms between the single parton and the full jet. Critically, the lost energy from the jet constituents may be partially brought back to the jet energy by the reconstruction procedures. This is an essential difference in energy loss mechanisms between the full-jet and the single parton, which leads to less energy loss of full-jet compared to heavy quarks. Therefore, stronger medium modification of $D(z_{||})$ can be obtained with larger R, which may be related to the R-dependence of jet energy loss [225,226]. Furthermore, stronger medium modification of $D(z_{||})$ is observed for B-jets compared to D-jets, due to their different initial spectra.

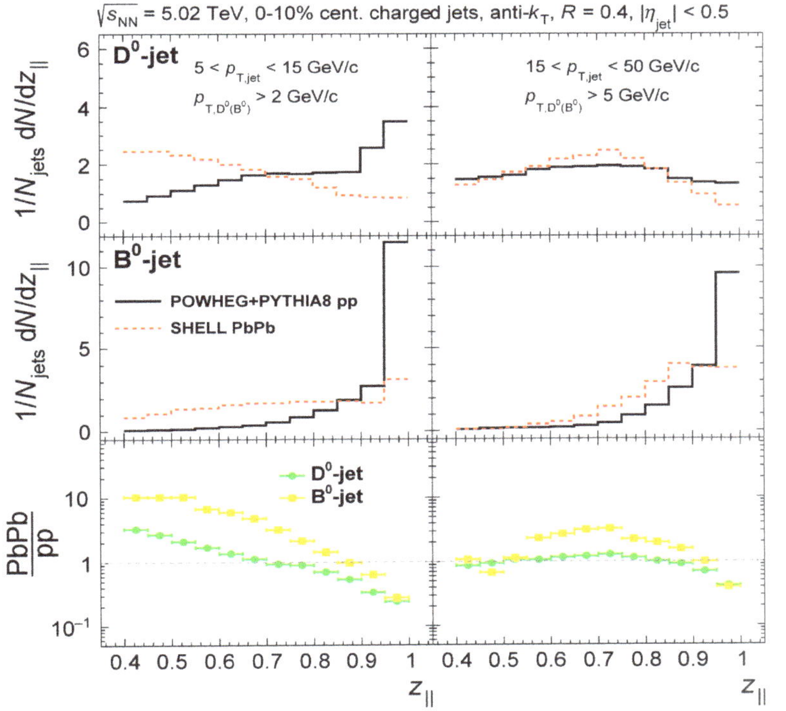

Figure 8. $D(z_{||})$ distributions of D-jet and B-jet within two p_T windows both in p+p and 0–10% Pb+Pb collisions, as well as the medium modifications (PbPb/pp). Figure is from Ref. [221].

3.6. The "Dead-Cone" Effect and Other Observables

Until now, there are a few other heavy-flavour jet observables accessible in the current experimental measurements at the LHC, which have also attracted attention from the high-energy nuclear physics community. We briefly discuss them in the following.

- The Cambridge-Aachen (CA) declustering techniques [227] which can help to obtain the angular-ordered pairwise tree of subjets [228] and the Soft Drop condition mentioned above enable us to expose the most basic heavy quark splitting structure by measuring the splitting-angle distributions in D^0 meson jets in p+p collisions at $\sqrt{s_{NN}} = 13$ TeV [229]. It has been measured in three different energy intervals of the radiators: $5 \le E_{\rm Radiator} \le 10$ GeV, $10 \le E_{\rm Radiator} \le 20$ GeV and

$20 \leq E_{\text{Radiator}} \leq 30$ GeV and constrain the transverse momentum of the D^0 meson in jet to be $2 < p_T^{D^0} < 36$ GeV/c. The ALICE collaboration directly observed for the first time a clear distribution suppression at the splitting angle smaller than the ratio of quark mass and the energy of such quark radiator: $\theta \leq M_{\text{charm}}/E_{\text{radiator}}$, known as the "dead-cone" effect [230,231]. Such a heavy quark jet and its substructure measurement reveal and confirm this most basic property of a fast quark interacting with the vacuum described by the QCD theory.

A subsequent phenomenology study exposed the "dead-cone" effect of the medium-induced gluon radiation of jet queching [53,150,184], by calculating the emission angle distribution of the heavy-flavour quark initiated splittings in a D^0 meson tagged jet and that of the light parton initiated splittings with the existence of the QGP in Pb+Pb collisions at $\sqrt{s_{NN}} = 5.02$ TeV [232], as demonstrated in Figure 9. Very interestingly, they find the collisional energy loss mechanism will not obscure the observation of the "dead-cone" effect in the medium-induced radiation. Such a proposal has also been verified by an analytical study that proposes a new jet substructure groomer that selects the most collinear splitting in a QCD jet above a certain transverse momentum cutoff [233]. It's also found in another study that the "dead-cone" domain would be partially filled by the medium-induced emission as heavy quarks traversing QGP [150].

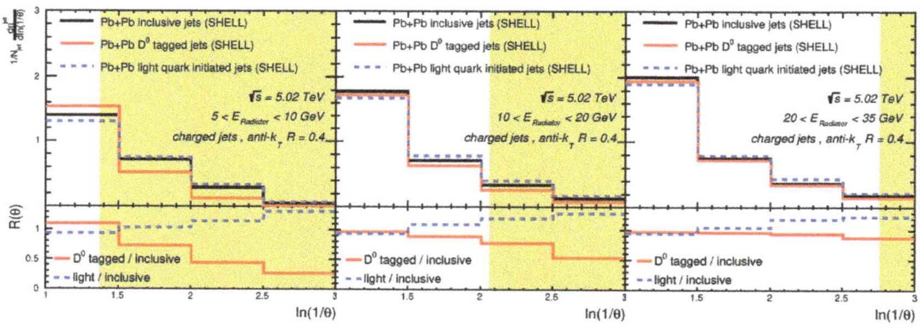

Figure 9. The splitting-angle distributions for D^0 meson tagged jets, inclusive jets and also light-quark jets normalized to the number of jets in Pb+Pb collisions at $\sqrt{s} = 5.02$ TeV (upper plots) and also the D^0 meson tagged jets/inclusive jets (light-quark jets/inclusive jets) ratios (bottom plots) calculated for three energy intervals of the radiators: $5 < E_{\text{Radiator}} < 10$ GeV (left panel), $10 < E_{\text{Radiator}} < 20$ GeV (middle panel) and $20 < E_{\text{Radiator}} < 30$ GeV (right panel). The shaded areas correspond to the angles at which the radiation is suppressed due to the "dead-cone" effect. Figure is from Ref. [232].

- The jet shape $\rho(r)$ describes the transverse energy profile of charged hadrons as a function of the angular distance from the jet axis. This observable has been well-studied for light flavor jets [205,234] to search the medium response effect as energetic parton dissipating energy to the medium [6]. The measurement of the medium modification on the b-jet shape has been reposted in Refs. [118,128] by the CMS collaboration. On the ond hand, the comparison of jet shapes of b-jets in Pb+Pb and p+p collisions shows the presence of the QGP modifies the energy distributions around the jet axis of b-jets. On the other hand, their measurements indicate a stronger jet energy redistribution of b-jets at larger radii compared to that of inclusive jets. Generally speaking, the bottom quarks are expected to dissipate less energy in nuclear collisions compared to light quarks and gluons due to the "dead-cone" effect. However, at larger jet radii, the medium response effect plays the dominant role in the enhancement of jet energy distribution in Pb+Pb collisions compared to the p+p baseline. Therefore, these interesting results may suggest that the heavier quark, like the bottom, may drive a stronger medium response effect than a massless parton. In this context, the heavy-flavour jets can serve as promising sensitive probes to the quasi-particle excitation of the quark soup.

- The Soft Drop (SD) grooming procedures reveal the two-prong structure of a jet, described by the momentum sharing z_g and opening angle R_g [228], which establishes the connection between the final state observable to the parton splitting function. The splitting history could be helpful to identify the production mechanisms of heavy-flavour jets [235,236]. Heavy quark jets from the gluon splitting process usually tend to have more balanced z_g and larger R_g compared to that from the FEX and FCR. The first measurement of the D-jet splitting function is performed by ALICE [237], and some theoretical efforts which focus on the medium modifications of z_g and R_g of c- and b-jets are presented in Refs. [238,239]. The medium effects result in more imbalanced z_g distribution and larger opening angles between the two subjets in the heavy quark jets, similar to the medium modification of inclusive jets observed by the CMS [240] and ALICE [241] collaboration.

4. Summary and Conclusions

This review covers the current development of theoretical studies on heavy-flavour jets in ultra-relativistic heavy-ion collisions. We introduce the recent theoretical advances of heavy-flavour production in heavy-ion collisions and then give a comprehensive discussion of several recent investigations relating to the heavy-flavour jet observables.

- We briefly overview the recent theoretical advances that help us understand the heavy-flavour production in heavy-ion collisions, mainly focusing on the initial production, transport approaches, hadronization mechanism, and diffusion coefficient extraction. These phenomenological studies based on the transport models reveal a fact that the elastic scattering of heavy quarks is dominant at lower p_T region ($p_T^Q < 5m_Q$), while the inelastic one dominate the high p_T regions. Besides, different from the fragmentation hadronization of heavy quarks in a vacuum, within the hot and dense nuclear matter, the coalescence mechanism plays an important role in explaining the large collective flow and the enhancement of baryon-to-meson ratio of a charmed hadron in nucleus–nucleus collisions at the RHIC and the LHC. The diffusion coefficient of heavy quarks in the QGP has been extracted by various theoretical frameworks, which implies that $2\pi TD_s$ slightly increases with temperature. The newly developed Bayesian inference approach may be promising to implement a robust determination of the transport coefficient of heavy quarks by a model-data fit.
- The studies on yield suppression and momentum imbalance of heavy-flavour jets are dedicated to addressing the mass effect of jet energy loss. Theoretical investigations predict stronger yield suppression of light quark jets compared to heavy-flavour jets, which is preliminarily proven by the recent ATLAS measurement of b-jet R_{AA}. However, the dijet asymmetry shows a reduced sensitivity to the jet quenching effect, therefore the difference of the medium modification on x_J between inclusive and $b\bar{b}$ dijets seems to be moderate. We have to say the nuclear modification factor is still an effective and powerful observable to test the mass effect of energy loss in QGP. On the other hand, the strategy to isolate the jets initiated by heavy quarks is also crucial to address the mass effect, since GSP processes indeed have a large contribution to the production of heavy quark jets but suffer stronger suppression in nucleus–nucleus collisions.
- An observable related to angular correlation aims at the deflection of the jet axis caused by the medium-induced p_T-broadening of jet quenching. It's found that the angular deviation caused by the in-medium scattering is hard to be observed for high-p_T jets, both for $b\bar{b}$ dijets and Z^0 + (b-)jet. That makes sense because higher p_T jets are more difficult to be changed by the in-medium scattering with the thermal parton in QGP. Meanwhile, medium modification on the radial profiles of jets containing lower-p_T D meson can well capture the angular de-correlation of the charm quark and the jet axis. This suggests that heavy flavors may be more suitable to address the medium-induced

p_T-broadening of jet quenching since they are experimentally accessible to the low-p_T domain where the angular deviation is visible.

- The substructure observable can reveal a wealth of information about the inner configuration of heavy-flavour jets. In the vacuum case, declustering techniques provide an inventive way to reestablish the splitting history of hard partons which helps us unlock the "dead-cone" effect of charm quark in the experiment. For heavy-flavour jets, the substructure observable also provides a unique opportunity to identify their production mechanisms. Furthermore, jet substructure, such as jet shape, seems more sensitive to the induced medium excitation in nucleus–nucleus collisions than full-jet observables. Much theoretical effort should be made to address the interplay of the "dead-cone" effect of medium-induced radiation and the medium response of heavy quarks. From the current perspective, the studies of substructures of heavy-flavour jets could play an increasingly important role in high-energy nuclear physics.
- The initial jet spectra and the "selection bias" play important roles in the medium modifications of jet substructure in nuclear collisions. Normally when we focus on the mass effect of the yield or substructure modification of heavy quark jets, it is apriori to believe that bottom jets should have a weaker medium modification in heavy-ion collisions compared to charm jets under the same conditions. However, in the studies of radial profile and fragmentation function of heavy-flavour jets, it's found that b-jets have very different initial substructure compared to that of c-jets event within the same kinematic constraints, which eventually leads to stronger medium modification of b-jets at the final-state compared to c-jets. On the other hand, the "selection bias" poses a challenge to the theoretical studies that aim at the nuclear modification mechanism of heavy-flavour jets in the hot and dense QCD medium. It brings additional "modifications" to the ratio PbPb/pp of jet substructure distributions, nevertheless, these "modifications" do not exactly reflect the change of jet substructure but only the decrease of jet energy from the higher kinematic region in Pb+Pb collisions.

Author Contributions: Conceptualization, S.W. and B.-W.Z.; methodology, W.D. and B.-W.Z.; investigation, S.W. and W.D.; writing—original draft preparation, S.W. and W.D.; writing—review and editing, B.-W.Z., E.W. and X.-N.W.; supervision, B.-W.Z., E.W. and X.-N.W. All authors have read and agreed to the published version of the manuscript.

Funding: This research is supported by the Guangdong Major Project of Basic and Applied Basic Research No. 2020B0301030008, and the Natural Science Foundation of China with Project Nos. 11935007, 12035007, 12247127. S. Wang is also supported by China Postdoctoral Science Foundation under project No. 2021M701279.

Data Availability Statement: Not applicable.

Conflicts of Interest: The authors declare no conflict of interest.

References

1. Wang, X.N.; Gyulassy, M. Gluon shadowing and jet quenching in A + A collisions at s**(1/2) = 200-GeV. *Phys. Rev. Lett.* **1992**, *68*, 1480–1483. [CrossRef] [PubMed]
2. Gyulassy, M.; Vitev, I.; Wang, X.N.; Zhang, B.W. Jet quenching and radiative energy loss in dense nuclear matter. *Quark Gluon Plasma* **2004**, *3*, 123–191.
3. Mehtar-Tani, Y.; Milhano, J.G.; Tywoniuk, K. Jet physics in heavy-ion collisions. *Int. J. Mod. Phys. A.* **2013**, *28*, 1340013. [CrossRef]
4. Qin, G.Y.; Wang, X.N. Jet quenching in high-energy heavy-ion collisions. *Int. J. Mod. Phys. E* **2015**, *24*, 1530014. [CrossRef]
5. Cunqueiro, L.; Sickles, A.M. Studying the QGP with Jets at the LHC and RHIC. *Prog. Part. Nucl. Phys.* **2022**, *124*, 103940. [CrossRef]
6. Cao, S.; Wang, X.N. Jet quenching and medium response in high-energy heavy-ion collisions: A review. *Rept. Prog. Phys.* **2021**, *84*, 024301. [CrossRef]
7. Collins, J.C.; Perry, M.J. Superdense Matter: Neutrons or Asymptotically Free Quarks? *Phys. Rev. Lett.* **1975**, *34*, 1353. [CrossRef]
8. Wang, X.N. Effect of jet quenching on high p_T hadron spectra in high-energy nuclear collisions. *Phys. Rev. C* **1998**, *58*, 2321. [CrossRef]
9. Wang, E.; Wang, X.N. Parton energy loss with detailed balance. *Phys. Rev. Lett.* **2001**, *87*, 142301. [CrossRef]

10. Majumder, A.; Wang, E.; Wang, X.N. Modified dihadron fragmentation functions in hot and nuclear matter. *Phys. Rev. Lett.* **2007**, *99*, 152301. [CrossRef]
11. Wang, E.; Wang, X.N. Jet tomography of dense and nuclear matter. *Phys. Rev. Lett.* **2002**, *89*, 162301. [CrossRef] [PubMed]
12. Zhang, H.; Owens, J.F.; Wang, E.; Wang, X.N. Dihadron tomography of high-energy nuclear collisions in NLO pQCD. *Phys. Rev. Lett.* **2007**, *98*, 212301. [CrossRef]
13. Vitev, I.; Zhang, B.W. Jet tomography of high-energy nucleus-nucleus collisions at next-to-leading order. *Phys. Rev. Lett.* **2010**, *104*, 132001. [CrossRef] [PubMed]
14. He, Y.; Pang, L.G.; Wang, X.N. Gradient Tomography of Jet Quenching in Heavy-Ion Collisions. *Phys. Rev. Lett.* **2020**, *125*, 122301. [CrossRef] [PubMed]
15. Wang, X.N.; Huang, Z.; Sarcevic, I. Jet quenching in the opposite direction of a tagged photon in high-energy heavy ion collisions. *Phys. Rev. Lett.* **1996**, *77*, 231–234. [CrossRef]
16. Neufeld, R.B.; Vitev, I.; Zhang, B.-W. The Physics of Z^0/γ^*-tagged jets at the LHC. *Phys. Rev. C* **2011**, *83*, 034902. [CrossRef]
17. Dai, W.; Vitev, I.; Zhang, B.W. Momentum imbalance of isolated photon-tagged jet production at RHIC and LHC. *Phys. Rev. Lett.* **2013**, *110*, 142001. [CrossRef]
18. Wang, X.N.; Zhu, Y. Medium Modification of γ-jets in High-energy Heavy-ion Collisions. *Phys. Rev. Lett.* **2013**, *111*, 062301. [CrossRef]
19. Li, H.; Liu, F.; Ma, G.l.; Wang, X.N.; Zhu, Y. Mach cone induced by γ-triggered jets in high-energy heavy-ion collisions. *Phys. Rev. Lett.* **2011**, *106*, 012301. [CrossRef]
20. Vitev, I.; Wicks, S.; Zhang, B.W. A Theory of jet shapes and cross-sections: From hadrons to nuclei. *JHEP* **2008**, *11*, 093. [CrossRef]
21. Caucal, P.; Soto-Ontoso, A.; Takacs, A. Dynamically groomed jet radius in heavy-ion collisions. *Phys. Rev. D* **2022**, *105*, 104046. [CrossRef]
22. Andronic, A.; Arleo, F.; Arnaldi, R.; Beraudo, A.; Bruna, E.; Caffarri, D.; Conesa del Valle, Z.; Contreras, J.G.; Dahms, T.; Dainese, A.; et al. Heavy-flavour and quarkonium production in the LHC era: From proton–proton to heavy-ion collisions. *Eur. Phys. J. C* **2016**, *76*, 107.
23. Zhang, B.W.; Ko, C.M.; Liu, W. Thermal charm production in a quark-gluon plasma in Pb-Pb collisions at s**(1/2)(NN) = 5.5-TeV. *Phys. Rev. C* **2008**, *77*, 024901. [CrossRef]
24. Adamczyk, L.; Adkins, J.K.; Agakishiev, G.; Aggarwal, M.M.; Ahammed, Z.; Alekseev, I.; Alford, J.; Anson, C.D.; Aparin, A.; Arkhipkin, D.; et al. Observation of D^0 Meson Nuclear Modifications in Au+Au Collisions at $\sqrt{s_{NN}} = 200$ GeV. *Phys. Rev. Lett.* **2014**, *113*, 142301. [CrossRef]
25. Adam, J.; Adamova, D.; Aggarwal, M.M.; Aglieri Rinella, G.; Agnello, M.; Agrawal, N.; Ahammed, Z.; Ahn, S.U.; Aiola, S.; Akindinov, A.; et al. Transverse momentum dependence of D-meson production in Pb-Pb collisions at $\sqrt{s_{NN}} = 2.76$ TeV. *JHEP* **2016**, *3*, 018.
26. Sirunyan, A.M.; Tumasyan, A.; Adam, W.; Ambrogi, F.; Asilar, E.; Bergauer, T.; Brandstetter, J.; Brondolin, E.; Dragicevic, M.; Erö, J.; et al. Nuclear modification factor of D^0 mesons in PbPb collisions at $\sqrt{s_{NN}} = 5.02$ TeV. *Phys. Lett. B* **2018**, *782*, 474. [CrossRef]
27. Adare, A.; Afanasiev, S.; Aidala, C.; Ajitanand, N.N.; Akiba, Y.; Al-Bataineh, H.; Alexander, J.; Aoki, K.; Aramaki, Y.; Atomssa, E.T.; et al. J/ψ suppression at forward rapidity in Au+Au collisions at $\sqrt{s_{NN}} = 200$ GeV. *Phys. Rev. C* **2011**, *84*, 054912. [CrossRef]
28. Adamczyk, L.; Adkins, J.K.; Agakishiev, G.; Aggarwal, M.M.; Ahammed, Z.; Alekseev, I.; Alford, J.; Anson, C.D.; Aparin, A.; Arkhipkin, D.; et al. J/ψ production at low p_T in Au + Au and Cu + Cu collisions at $\sqrt{s_{NN}} = 200$ GeV with the STAR detector. *Phys. Rev. C* **2014**, *90*, 024906. [CrossRef]
29. Abelev, B.B.; Adam, J.; Adamova, D.; Mohan Aggarwal, M.; Aglieri Rinella, G.; Agnello, M.; Agostinelli, A.; Agrawal, N.; Ahammed, Z.; Ahmad, N.; et al. Suppression of $\Upsilon(1S)$ at forward rapidity in Pb-Pb collisions at $\sqrt{s_{NN}} = 2.76$ TeV. *Phys. Lett. B* **2014**, *738*, 361–372. [CrossRef]
30. Abelev, B.B.; Adam, J.; Adamova, D.; Aggarwal, M.M.; Aglieri Rinella, G.; Agnello, M.; Agostinelli, A.; Agrawal, N.; Ahammed, Z.; Ahmad, N.; et al. Indications of suppression of excited Υ states in PbPb collisions at $\sqrt{s_{NN}} = 2.76$ TeV. *Phys. Rev. Lett.* **2011**, *107*, 052302.
31. Adam, J.; Adamczyk, L.; Adams, J.R.; Adkins, J.K.; Agakishiev, G.; Aggarwal, M.M.; Ahammed, Z.; Alekseev, I.; Anderson, D.M.; Aoyama, R.; et al. First Observation of the Directed Flow of D^0 and $\overline{D^0}$ in Au+Au Collisions at $\sqrt{s_{NN}} = 200$ GeV. *Phys. Rev. Lett.* **2019**, *123*, 162301. [CrossRef]
32. Acharya, S.; Adamova, D.; Adler, A.; Adolfsson, J.; Aggarwal, M.M.; Aglieri Rinella, G.; Agnello, M.; Agrawal, N.; Ahammed, Z.; Ahmad, S.; et al. Probing the effects of strong electromagnetic fields with charge-dependent directed flow in Pb-Pb collisions at the LHC. *Phys. Rev. Lett.* **2020**, *125*, 022301. [CrossRef]
33. Abelev, B.B.; Adam, J.; Adamova, D.; Mohan Aggarwal, M.; Aglieri Rinella, G.; Agnello, M.; Agostinelli, A.; Agrawal, N.; Ahammed, Z.; Ahmad, N.; et al. Azimuthal anisotropy of D meson production in Pb-Pb collisions at $\sqrt{s_{NN}} = 2.76$ TeV. *Phys. Rev. C* **2014**, *90*, 034904. [CrossRef]
34. Adamczyk, L.; Adkins, J.K.; Agakishiev, G.; Aggarwal, M.M.; Ahammed, Z.; Ajitanand, N.N.; Alekseev, I.; Anderson, D.M.; Aoyama, R.; Aparin, A.; et al. Measurement of D^0 Azimuthal Anisotropy at Midrapidity in Au+Au Collisions at $\sqrt{s_{NN}} = 200$ GeV. *Phys. Rev. Lett.* **2017**, *118*, 212301. [CrossRef] [PubMed]

35. Acharya, S.; Adamova, D.; Adolfsson, J.; Aggarwal, M.M.; Aglieri Rinella, G.; Agnello, M.; Agrawal, N.; Ahammed, Z.; Ahmad, S.; Ahmad, N.; et al. *D*-meson azimuthal anisotropy in midcentral Pb-Pb collisions at $\sqrt{s_{NN}} = 5.02$ TeV. *Phys. Rev. Lett.* **2018**, *120*, 102301. [CrossRef] [PubMed]
36. Sirunyan, A.M.; Tumasyan, A.; Adam, W.; Ambrogi, F.; Asilar, E.; Bergauer, T.; Brandstetter, J.; Brondolin, E.; Dragicevic, M.; Erö, J.; et al. Measurement of prompt D^0 meson azimuthal anisotropy in Pb-Pb collisions at $\sqrt{s_{NN}} = 5.02$ TeV. *Phys. Rev. Lett.* **2018**, *120*, 202301. [CrossRef] [PubMed]
37. van Hees, H.; Mannarelli, M.; Greco, V.; Rapp, R. Nonperturbative heavy-quark diffusion in the quark-gluon plasma. *Phys. Rev. Lett.* **2008**, *100*, 192301. [CrossRef]
38. Caron-Huot, S.; Moore, G.D. Heavy quark diffusion in QCD and N = 4 SYM at next-to-leading order. *JHEP* **2008**, *02*, 081. [CrossRef]
39. Djordjevic, M.; Djordjevic, M. Predictions of heavy-flavor suppression at 5.1 TeV Pb + Pb collisions at the CERN Large Hadron Collider. *Phys. Rev. C* **2015**, *92*, 024918. [CrossRef]
40. He, M.; Fries, R.J.; Rapp, R. Heavy Flavor at the Large Hadron Collider in a Strong Coupling Approach. *Phys. Lett. B* **2014**, *735*, 445. [CrossRef]
41. Kang, Z.B.; Ringer, F.; Vitev, I. Effective field theory approach to open heavy flavor production in heavy-ion collisions. *JHEP* **2017**, *3*, 146. [CrossRef]
42. Svetitsky, B. Diffusion of charmed quarks in the quark-gluon plasma. *Phys. Rev. D* **1988**, *37*, 2484. [CrossRef] [PubMed]
43. Moore, G.D.; Teaney, D. How much do heavy quarks thermalize in a heavy ion collision? *Phys. Rev. C* **2005**, *71*, 064904. [CrossRef]
44. Cao, S.; Qin, G.Y.; Bass, S.A. Heavy-quark dynamics and hadronization in ultrarelativistic heavy-ion collisions: Collisional versus radiative energy loss. *Phys. Rev. C* **2013**, *88*, 044907. [CrossRef]
45. Alberico, W.M.; Beraudo, A.; De Pace, A.; Molinari, A.; Monteno, M.; Nardi, M.; Prino, F.; Sitta, M. Heavy flavors in *AA* collisions: Production, transport and final spectra. *Eur. Phys. J. C* **2018**, *73*, 2481. [CrossRef]
46. Xu, J.; Liao, J.; Gyulassy, M. Bridging Soft-Hard Transport Properties of Quark-Gluon Plasmas with CUJET3.0. *JHEP* **2016**, *2*, 169. [CrossRef]
47. Cao, S.; Luo, T.; Qin, G.Y.; Wang, X.N. Linearized Boltzmann transport model for jet propagation in the quark-gluon plasma: Heavy quark evolution. *Phys. Rev. C* **2016**, *94*, 014909. [CrossRef]
48. Das, S.K.; Plumari, S.; Chatterjee, S.; Alam, J.; Scardina, F.; Greco, V. Directed Flow of Charm Quarks as a Witness of the Initial Strong Magnetic Field in Ultra-Relativistic Heavy Ion Collisions. *Phys. Lett. B* **2017**, *768*, 260. [CrossRef]
49. Ke, W.; Xu, Y.; Bass, S.A. Linearized Boltzmann-Langevin model for heavy quark transport in hot and dense QCD matter. *Phys. Rev. C* **2018**, *98*, 064901. [CrossRef]
50. Li, S.Q.; Xing, W.J.; Liu, F.L.; Cao, S.; Qin, G.Y. Heavy flavor quenching and flow: The roles of initial condition, pre-equilibrium evolution, and in-medium interaction. *Chin. Phys. C* **2020**, *44*, 114101. [CrossRef]
51. Yan, L.; Zhuang, P.; Xu, N. Competition between J/psi suppression and regeneration in quark-gluon plasma. *Phys. Rev. Lett.* **2006**, *97*, 232301. [CrossRef] [PubMed]
52. Liu, Y.; Qu, Z.; Xu, N.; Zhuang, P. J/psi Transverse Momentum Distribution in High Energy Nuclear Collisions at RHIC. *Phys. Lett. B* **2009**, *678*, 72–76. [CrossRef]
53. Zhang, B.W.; Wang, E.; Wang, X.N. Heavy quark energy loss in nuclear medium. *Phys. Rev. Lett.* **2004**, *93*, 072301. [CrossRef] [PubMed]
54. Djordjevic, M.; Gyulassy, M. Heavy quark radiative energy loss in QCD matter. *Nucl. Phys. A* **2004**, *733*, 265–298. [CrossRef]
55. Zhang, B.W.; Wang, E.; Wang, X.N. Multiple parton scattering in nuclei: Heavy quark energy loss and modified fragmentation functions. *Nucl. Phys. A* **2005**, *757*, 493–524. [CrossRef]
56. Fan, W.; Vujanovic, G.; Bass, S.A.; Majumder, A.; Angerami, A.; Arora, R.; Cao, S.; Chen, Y.; Dai, T.; Du, L.; et al. Multi-scale evolution of charmed particles in a nuclear medium. *arXiv* **2022**, arXiv:2208.00983.
57. Liu, F.L.; Xing, W.J.; Wu, X.Y.; Qin, G.Y.; Cao, S.; Wang, X.N. QLBT: A linear Boltzmann transport model for heavy quarks in a quark-gluon plasma of quasi-particles. *Eur. Phys. J. C* **2022**, *82*, 350. [CrossRef]
58. Ding, C.; Ke, W.Y.; Pang, L.G.; Wang, X.N. Hydrodynamic description of D meson production in high-energy heavy-ion collisions. *Chin. Phys. C* **2021**, *45*, 074102. [CrossRef]
59. Cao, S.; Sun, K.J.; Li, S.Q.; Liu, S.Y.F.; Xing, W.J.; Qin, G.Y.; Ko, C.M. Charmed hadron chemistry in relativistic heavy-ion collisions. *Phys. Lett. B* **2020**, *807*, 135561. [CrossRef]
60. Plumari, S.; Minissale, V.; Das, S.K.; Coci, G.; Greco, V. Charmed Hadrons from Coalescence plus Fragmentation in relativistic nucleus-nucleus collisions at RHIC and LHC. *Eur. Phys. J. C* **2018**, *78*, 348. [CrossRef]
61. He, M.; Rapp, R. Hadronization and Charm-Hadron Ratios in Heavy-Ion Collisions. *Phys. Rev. Lett.* **2020**, *124*, 042301. [CrossRef]
62. Rapp, R.; van Hees, H. Heavy Quarks in the Quark-Gluon Plasma. *arXiv* **2009**, arXiv:0903.1096.
63. Prino, F.; Rapp, R. Open Heavy Flavor in QCD Matter and in Nuclear Collisions. *J. Phys. G* **2016**, *43*, 093002. [CrossRef]
64. Rapp, R.; Gossiaux, P.B.; Andronic, A.; Averbeck, R.; Masciocchi, S.; Beraudo, A.; Bratkovskaya, E.; Braun-Munzinger, P.; Cao, S.; Dainese, A.; et al. Extraction of Heavy-Flavor Transport Coefficients in QCD Matter. *Nucl. Phys. A* **2018**, *979*, 21–86.
65. Dong, X.; Greco, V. Heavy quark production and properties of Quark–Gluon Plasma. *Prog. Part. Nucl. Phys.* **2019**, *104*, 97. [CrossRef]

66. Dong, X.; Lee, Y.J.; Rapp, R. Open Heavy-Flavor Production in Heavy-Ion Collisions. *Ann. Rev. Nucl. Part. Sci.* **2019**, *69*, 417–445. [CrossRef]
67. Zhao, J.; Zhou, K.; Chen, S.; Zhuang, P. Heavy flavors under extreme conditions in high energy nuclear collisions. *Prog. Part. Nucl. Phys.* **2020**, *114*, 103801. [CrossRef]
68. Apolinário, L.; Lee, Y.J.; Winn, M. Heavy quarks and jets as probes of the QGP. *Prog. Part. Nucl. Phys.* **2022**, *127*, 103990. [CrossRef]
69. He, M.; van Hees, H.; Rapp, R. Heavy-Quark Diffusion in the Quark-Gluon Plasma. *arXiv* **2022**, arXiv:2204.09299.
70. Tang, Z.; Tang, Z.B.; Zha, W.; Zha, W.M.; Zhang, Y.; Zhang, Y.F. An experimental review of open heavy flavor and quarkonium production at RHIC. *Nucl. Sci. Tech.* **2020**, *31*, 81. [CrossRef]
71. Li, H.T.; Vitev, I. Inclusive heavy flavor jet production with semi-inclusive jet functions: From proton to heavy-ion collisions. *JHEP* **2019**, *7*, 148. [CrossRef]
72. Xu, J.; Liao, J.; Gyulassy, M. Consistency of Perfect Fluidity and Jet Quenching in semi-Quark-Gluon Monopole Plasmas. *Chin. Phys. Lett.* **2015**, *32*, 092501. [CrossRef]
73. Djordjevic, M. Theoretical formalism of radiative jet energy loss in a finite size dynamical QCD medium. *Phys. Rev. C* **2009**, *80*, 064909. [CrossRef]
74. Djordjevic, M.; Heinz, U.W. Radiative energy loss in a finite dynamical QCD medium. *Phys. Rev. Lett.* **2008**, *101*, 022302. [CrossRef]
75. Djordjevic, M.; Djordjevic, M. LHC jet suppression of light and heavy flavor observables. *Phys. Lett. B* **2014**, *734*, 286–289. [CrossRef]
76. Zigic, D.; Salom, I.; Auvinen, J.; Huovinen, P.; Djordjevic, M. DREENA-A framework as a QGP tomography tool. *Front. in Phys.* **2022**, *10*, 957019. [CrossRef]
77. Zigic, D.; Salom, I.; Auvinen, J.; Djordjevic, M.; Djordjevic, M. DREENA-B framework: First predictions of R_{AA} and v_2 within dynamical energy loss formalism in evolving QCD medium. *Phys. Lett. B* **2019**, *791*, 236–241. [CrossRef]
78. Zigic, D.; Salom, I.; Auvinen, J.; Djordjevic, M.; Djordjevic, M. DREENA-C framework: Joint R_{AA} and v_2 predictions and implications to QGP tomography. *J. Phys. G* **2019**, *46*, 085101. [CrossRef]
79. Wicks, S.; Horowitz, W.; Djordjevic, M.; Gyulassy, M. Heavy quark jet quenching with collisional plus radiative energy loss and path length fluctuations. *Nucl. Phys. A* **2007**, *783*, 493–496. [CrossRef]
80. Wicks, S.; Horowitz, W.; Djordjevic, M.; Gyulassy, M. Elastic, inelastic, and path length fluctuations in jet tomography. *Nucl. Phys. A* **2007**, *784*, 426–442. [CrossRef]
81. Horowitz, W.A.; Gyulassy, M. Heavy quark jet tomography of Pb + Pb at LHC: AdS/CFT drag or pQCD energy loss? *Phys. Lett. B* **2008**, *666*, 320–323. [CrossRef]
82. Horowitz, W.A. Testing pQCD and AdS/CFT Energy Loss at RHIC and LHC. *AIP Conf. Proc.* **2012**, *1441*, 889–891.
83. Uphoff, J.; Fochler, O.; Xu, Z.; Greiner, C. Heavy quark production at RHIC and LHC within a partonic transport model. *Phys. Rev. C* **2010**, *82*, 044906. [CrossRef]
84. Uphoff, J.; Fochler, O.; Xu, Z.; Greiner, C. Elliptic Flow and Energy Loss of Heavy Quarks in Ultra-Relativistic heavy Ion Collisions. *Phys. Rev. C* **2011**, *84*, 024908. [CrossRef]
85. Uphoff, J.; Senzel, F.; Xu, Z.; Greiner, C. Momentum imbalance of D mesons in ultra-relativistic heavy-ion collisions at LHC. *Phys. Rev. C* **2014**, *89*, 064906. [CrossRef]
86. Uphoff, J.; Fochler, O.; Xu, Z.; Greiner, C. Elastic and radiative heavy quark interactions in ultra-relativistic heavy-ion collisions. *J. Phys. G* **2015**, *42*, 115106. [CrossRef]
87. Gossiaux, P.B.; Aichelin, J. Towards an understanding of the RHIC single electron data. *Phys. Rev. C* **2008**, *78*, 014904. [CrossRef]
88. Gossiaux, P.B.; Bierkandt, R.; Aichelin, J. Tomography of a quark gluon plasma at RHIC and LHC energies. *Phys. Rev. C* **2009**, *79*, 044906. [CrossRef]
89. Nahrgang, M.; Aichelin, J.; Gossiaux, P.B.; Werner, K. Azimuthal correlations of heavy quarks in Pb + Pb collisions at $\sqrt{s} = 2.76$ TeV at the CERN Large Hadron Collider. *Phys. Rev. C* **2014**, *90*, 024907. [CrossRef]
90. Ke, W.; Wang, X.N. QGP modification to single inclusive jets in a calibrated transport model. *JHEP* **2021**, *05*, 041. [CrossRef]
91. Plumari, S.; Puglisi, A.; Scardina, F.; Greco, V. Shear Viscosity of a strongly interacting system: Green-Kubo vs. Chapman-Enskog and Relaxation Time Approximation. *Phys. Rev. C* **2012**, *86*, 054902. [CrossRef]
92. Das, S.K.; Scardina, F.; Plumari, S.; Greco, V. Toward a solution to the R_{AA} and v_2 puzzle for heavy quarks. *Phys. Lett. B* **2015**, *747*, 260–264. [CrossRef]
93. Plumari, S.; Alberico, W.M.; Greco, V.; Ratti, C. Recent thermodynamic results from lattice QCD analyzed within a quasi-particle model. *Phys. Rev. D* **2011**, *84*, 094004. [CrossRef]
94. Scardina, F.; Das, S.K.; Minissale, V.; Plumari, S.; Greco, V. Estimating the charm quark diffusion coefficient and thermalization time from D meson spectra at energies available at the BNL Relativistic Heavy Ion Collider and the CERN Large Hadron Collider. *Phys. Rev. C* **2017**, *96*, 044905. [CrossRef]
95. Alberico, W.M.; Beraudo, A.; De Pace, A.; Molinari, A.; Monteno, M.; Nardi, M.; Prino, F. Heavy-flavour spectra in high energy nucleus-nucleus collisions. *Eur. Phys. J. C* **2011**, *71*, 1666. [CrossRef]
96. Beraudo, A.; Pace, A.D.; Monteno, M.; Nardi, M.; Prino, F. Heavy flavors in heavy-ion collisions: Quenching, flow and correlations. *Eur. Phys. J. C* **2015**, *75*, 121. [CrossRef]

97. Cao, S.; Bass, S.A. Thermalization of charm quarks in infinite and finite QGP matter. *Phys. Rev. C* **2011**, *84*, 064902. [CrossRef]
98. Lang, T.; van Hees, H.; Steinheimer, J.; Inghirami, G.; Bleicher, M. Heavy quark transport in heavy ion collisions at energies available at the BNL Relativistic Heavy Ion Collider and at the CERN Large Hadron Collider within the UrQMD hybrid model. *Phys. Rev. C* **2016**, *93*, 014901. [CrossRef]
99. Lang, T.; van Hees, H.; Steinheimer, J.; Bleicher, M. Elliptic flow and nuclear modification factors of D-mesons at FAIR in a Hybrid-Langevin approach. *arXiv* **2013**, arXiv:1305.1797.
100. Lang, T.; van Hees, H.; Steinheimer, J.; Bleicher, M. Dileptons from correlated D- and $\bar{D}$-meson decays in the invariant mass range of the QGP thermal radiation using the UrQMD hybrid model. *arXiv* **2013**, arXiv:1305.7377.
101. He, M.; Fries, R.J.; Rapp, R. $\mathbf{D_s}$-Meson as Quantitative Probe of Diffusion and Hadronization in Nuclear Collisions. *Phys. Rev. Lett.* **2013**, *110*, 112301. [CrossRef]
102. He, M.; Fries, R.J.; Rapp, R. Thermal Relaxation of Charm in Hadronic Matter. *Phys. Lett. B* **2011**, *701*, 445–450. [CrossRef]
103. Wang, S.; Dai, W.; Zhang, B.W.; Wang, E. Diffusion of charm quarks in jets in high-energy heavy-ion collisions. *Eur. Phys. J. C* **2019**, *79*, 789. [CrossRef]
104. Dai, W.; Wang, S.; Zhang, B.W.; Wang, E. Transverse Momentum Balance and Angular Distribution of $b\bar{b}$ Dijets in Pb+Pb collisions. *Chin. Phys. C* **2020**, *44*, 104105. [CrossRef]
105. Wang, S.; Kang, J.W.; Dai, W.; Zhang, B.W.; Wang, E. Probing the in-medium P_T-broadening by γ+HF angular de-correlations. *Eur. Phys. J. A* **2022**, *58*, 135. [CrossRef]
106. Cassing, W.; Bratkovskaya, E.L. Parton transport and hadronization from the dynamical quasiparticle point of view. *Phys. Rev. C* **2008**, *78*, 034919. [CrossRef]
107. Cassing, W.; Bratkovskaya, E.L. Parton-Hadron-String Dynamics: An off-shell transport approach for relativistic energies. *Nucl. Phys. A* **2009**, *831*, 215–242. [CrossRef]
108. Bratkovskaya, E.L.; Cassing, W.; Konchakovski, V.P.; Linnyk, O. Parton-Hadron-String Dynamics at Relativistic Collider Energies. *Nucl. Phys. A* **2011**, *856*, 162–182. [CrossRef]
109. Adam, J.; Adamczyk, L.; Adams, J.R.; Adkins, J.K.; Agakishiev, G.; Aggarwal, M.M.; Ahammed, Z.; Alekseev, I.; Anderson, D.M.; Aoyama, R.; et al. First measurement of Λ_c baryon production in Au+Au collisions at $\sqrt{s_{NN}}$ = 200 GeV. *Phys. Rev. Lett.* **2020**, *124*, 172301. [CrossRef]
110. Vermunt, L. Measurement of Λ_c baryons and D_s^+ mesons in Pb–Pb collisions with ALICE. *Pos* **2020**, *EPS-HEP2019*, 297.
111. Sirunyan, A.M.; Tumasyan, A.; Adam, W.; Asilar, E.; Bergauer, T.; Brandstetter, J.; Brondolin, E.; Dragicevic, M.; Erö, J.; Dragicevic, M.; et al. Measurements of the charm jet cross-section and nuclear modification factor in pPb collisions at $\sqrt{s_{NN}}$ = 5.02 TeV. *Phys. Lett. B* **2017**, *772*, 306–329. [CrossRef]
112. Acharya, S.; Adamova, D.; Adler, A.; Adolfsson, J.; Aggarwal, M.M.; Aglieri Rinella, G.; Agnello, M.; Agrawal, N.; Ahammed, Z.; Ahmad, S.; et al. Measurement of the production of charm jets tagged with D^0 mesons in pp collisions at $\sqrt{s} = 7$ TeV. *JHEP* **2019**, *8*, 133.
113. Chatrchyan, S.; Khachatryan, V.; Sirunyan, A.M.; Tumasyan, A.; Adam, W.; Bergauer, T.; Dragicevic, M.; Erö, J.; Fabjan, C.; Friedl, M.; et al. Inclusive b-jet production in pp collisions at $\sqrt{s} = 7$ TeV. *JHEP* **2012**, *4*, 084.
114. Aad, G.; Abbott, B.; Abdallah, J.; Abdelalim, A.A.; Abdesselam, A.; Abdinov, O.; Abi, B.; Abolins, M.; Abramowicz, H.; Abreu, H.; et al. Measurement of $D^{*+/-}$ meson production in jets from pp collisions at sqrt(s) = 7 TeV with the ATLAS detector. *Phys. Rev. D* **2012**, *85*, 052005. [CrossRef]
115. Aad, G.; Abbott, B.K.; Abbott, D.; Abud, A.A.; Abeling, K.; Abhayasinghe, D.K.; Abidi, H.; AbouZeid, O.S.; Abramowicz, H.; Abreu, H.; et al. Measurement of b-quark fragmentation properties in jets using the decay $B^{\pm} \to J/\psi K^{\pm}$ in pp collisions at $\sqrt{s}$ = 13 TeV with the ATLAS detector. *JHEP* **2012**, *12*, 131.
116. Abelev, B.I.; Aggarwal, M.M.; Ahammed, Z.; Anderson, B.D.; Arkhipkin, D.; Averichev, G.S.; Balewski, J.; Barannikova, O.; Barnby, L.S.; Baudot, J.; et al. Measurement of D* Mesons in Jets from p+p Collisions at s**(1/2) = 200-GeV. *Phys. Rev. D* **2009**, *79*, 112006. [CrossRef]
117. Aaboud, M.; Aad, G.; Abbott, B.; Abdallah, J.; Abdinov, O.; Abeloos, B.; Aben, R.; AbouZeid, O.; Abraham, N.; Abramowicz, H.; et al. Measurement of the $b\bar{b}$ dijet cross-section in pp collisions at $\sqrt{s} = 7$ TeV with the ATLAS detector. *Eur. Phys. J. C* **2016**, *76*, 670. [CrossRef] [PubMed]
118. Sirunyan, A.M.; Tumasyan, A.; Adam, W.; Ambrogi, F.; Bergauer, T.; Dragicevic, M.; Erö, J.; Del Valle, A.E.; Flechl, M.; Fruehwirth, R.; et al. Measurement of b jet shapes in proton-proton collisions at $\sqrt{s} = 5.02$ TeV. *JHEP* **2021**, *5*, 054.
119. Khachatryan, V.; Sirunyan, A.M.; Tumasyan, A.; Adam, W.; Asilar, E.; Bergauer, T.; Brandstetter, J.; Brondolin, E.; Dragicevic, M.; Erö, J.; et al. Transverse momentum spectra of inclusive b jets in pPb collisions at $\sqrt{s_{NN}} = 5.02$ TeV. *Phys. Lett. B* **2016**, *754*, 59. [CrossRef]
120. Acharya, S.; Adamova, D.; Adler, A.; Adolfsson, J.; Aggarwal, M.M.; Aglieri Rinella, G.; Agnello, M.; Agrawal, N.; Ahammed, Z.; Ahmad, S.; et al. Measurement of inclusive charged-particle b-jet production in pp and p-Pb collisions at $\sqrt{s_{NN}}$ = 5.02 TeV. *JHEP* **2022**, *1*, 178.
121. Chatrchyan, S.; Khachatryan, V.; Sirunyan, A.M.; Tumasyan, A.; Adam, W.; Bergauer, T.; Dragicevic, M.; Erö, J.; Fabjan, C.; Friedl, M.; et al. Evidence of b-Jet Quenching in PbPb Collisions at $\sqrt{s_{NN}} = 2.76$ TeV. *Phys. Rev. Lett.* **2014**, *113*, 132301. [CrossRef] [PubMed]

122. Sirunyan, A.M.; Tumasyan, A.; Adam, W.; Asilar, E.; Bergauer, T.; Brandstetter, J.; Brondolin, E.; Dragicevic, M.; Erö, J.; Dragicevic, M.; et al. Comparing transverse momentum balance of b jet pairs in pp and PbPb collisions at $\sqrt{s_{NN}} = 5.02$ TeV. *JHEP* **2018**, *3*, 181.
123. Sirunyan, A.M.; Tumasyan, A.; Adam, W.; Asilar, E.; Bergauer, T.; Brandstetter, J.; Brondolin, E.; Dragicevic, M.; Erö, J.; Dragicevic, M.; et al. Studies of charm quark diffusion inside jets using PbPb and pp collisions at $\sqrt{s_{NN}} = 5.02$ TeV. *Phys. Rev. Lett.* **2019**, *125*, 102001. [CrossRef] [PubMed]
124. ATLAS Collaboration. Measurement of the nuclear modification factor of *b*-jets in 5.02 TeV Pb+Pb collisions with the ATLAS detector. *arXiv* **2022**, arXiv:2204.13530.
125. Roy, D. An Investigation of Charm Quark Jet Spectrum and Shape Modifications in Au+Au Collisions at $\sqrt{s_{NN}} = 200$ GeV. *arXiv* **2022**, arXiv:2207.14434.
126. Tumasyan, A.; Adam, W.; Bergauer, T.; Dragicevic, M.; Erö, J.; Valle, A.E.D.; Fruehwirth, R.; Jeitler, M.; Krammer, N.; Lechner, L.; et al. Fragmentation of jets containing a prompt J/ψ meson in PbPb and pp collisions at $\sqrt{s_{NN}} = 5.02$ TeV. *Phys. Lett. B* **2022**, *825*, 136842. [CrossRef]
127. Acharya, S.; Torales-Acosta, F.; Adamova, D.; Adolfsson, J.; Aggarwal, M.M.; Rinella, G.A.; Agnello, M.; Agrawal, N.; Ahammed, Z.; Ahn, S.U.; et al. Measurement of D^0, D^+, D^{*+} and D_s^+ production in Pb-Pb collisions at $\sqrt{s_{NN}} = 5.02$ TeV. *JHEP* **2018**, *10*, 174.
128. CMS Collaboration. Search for medium effects using jets from bottom quarks in PbPb collisions at $\sqrt{s_{NN}} = 5.02$ TeV. *arXiv* **2022**, arXiv:2210.08547.
129. Cacciari, M.; Nason, P.; Vogt, R. QCD predictions for charm and bottom production at RHIC. *Phys. Rev. Lett.* **2005**, *95*, 122001. [CrossRef]
130. Eskola, K.J.; Paukkunen, H.; Salgado, C.A. EPS09: A New Generation of NLO and LO Nuclear Parton Distribution Functions. *JHEP* **2009**, *4*, 065. [CrossRef]
131. Eskola, K.J.; Paakkinen, P.; Paukkunen, H.; Salgado, C.A. EPPS16: Nuclear parton distributions with LHC data. *Eur. Phys. J. C* **2017**, *77*, 163. [CrossRef] [PubMed]
132. Ball, R.D.; Bertone, V.; Carrazza, S.; Deans, C.S.; Debbio, L.D.; Forte, S.; Guffanti, A.; Hartland, N.P.; Latorre, J.I.; Rojo, J.; et al. Parton distributions for the LHC Run II. *JHEP* **2015**, *4*, 040. [CrossRef]
133. Peterson, C.; Schlatter, D.; Schmitt, I.; Zerwas, P.M. Scaling Violations in Inclusive e+ e- Annihilation Spectra. *Phys. Rev. D* **1983**, *27*, 105. [CrossRef]
134. Andersson, B.; Gustafson, G.; Ingelman, G.; Sjostrand, T. Parton Fragmentation and String Dynamics. *Phys. Rept.* **1983**, *97*, 31–145. [CrossRef]
135. Aversa, F.; Chiappetta, P.; Greco, M.; Guillet, J.P. QCD Corrections to Parton-Parton Scattering Processes. *Nucl. Phys. B* **1989**, *327*, 105. [CrossRef]
136. Kniehl, B.A.; Kramer, G.; Schienbein, I.; Spiesberger, H. Inclusive D^{*+-} production in p anti-p collisions with massive charm quarks. *Phys. Rev. D* **2005**, *71*, 014018. [CrossRef]
137. Kniehl, B.A.; Kramer, G.; Schienbein, I.; Spiesberger, H. Collinear subtractions in hadroproduction of heavy quarks. *Eur. Phys. J. C* **2005**, *41*, 199–212. [CrossRef]
138. Cacciari, M.; Frixione, S.; Houdeau, N.; Mangano, M.L.; Nason, P.; Ridolfi, G. Theoretical predictions for charm and bottom production at the LHC. *JHEP* **2012**, *1210*, 137. [CrossRef]
139. Sjostrand, T.; Mrenna, S.; Skands, P.Z. PYTHIA 6.4 Physics and Manual. *JHEP* **2006**, *05*, 026. [CrossRef]
140. Corcella, G.; Knowles, I.G.; Marchesini, G.; Moretti, S.; Odagiri, K.; Richardson, P.; Seymour, M.H.; Webber, B.R. HERWIG 6: An Event generator for hadron emission reactions with interfering gluons (including supersymmetric processes. *JHEP* **2001**, *01*, 010. [CrossRef]
141. Frixione, S.; Nason, P.; Oleari, C. Matching NLO QCD computations with Parton Shower simulations: The POWHEG method. *JHEP* **2007**, *11*, 070. [CrossRef]
142. Gleisberg, T.; Hoeche, S.; Krauss, F.; Schonherr, M.; Schumann, S.; Siegert, F.; Winter, J. Event generation with SHERPA 1.1. *JHEP* **2009**, *2*, 007. [CrossRef]
143. Ferini, G.; Colonna, M.; Di Toro, M.; Greco, V. Scalings of Elliptic Flow for a Fluid at Finite Shear Viscosity. *Phys. Lett. B* **2009**, *670*, 325–329. [CrossRef]
144. Ruggieri, M.; Scardina, F.; Plumari, S.; Greco, V. Thermalization, Isotropization and Elliptic Flow from Nonequilibrium Initial Conditions with a Saturation Scale. *Phys. Rev. C* **2014**, *89*, 054914. [CrossRef]
145. Beraudo, A. Dynamics of heavy flavor quarks in high energy nuclear collisions. *Nucl. Phys. A* **2014**, *931*, 145–154. [CrossRef]
146. Arnold, P.B.; Moore, G.D.; Yaffe, L.G. Photon and gluon emission in relativistic plasmas. *JHEP* **2002**, *06*, 030. [CrossRef]
147. Abir, R.; Greiner, C.; Martinez, M.; Mustafa, M.G.; Uphoff, J. Soft gluon emission off a heavy quark revisited. *Phys. Rev. D* **2012**, *85*, 054012. [CrossRef]
148. Abir, R.; Jamil, U.; Mustafa, M.G.; Srivastava, D.K. Heavy quark energy loss and D-mesons in RHIC and LHC energies. *Phys. Lett. B* **2012**, *715*, 183–189. [CrossRef]
149. Zapp, K.; Ingelman, G.; Rathsman, J.; Stachel, J.; Wiedemann, U.A. A Monte Carlo Model for 'Jet Quenching'. *Eur. Phys. J. C* **2009**, *60*, 617–632. [CrossRef]

150. Armesto, N.; Salgado, C.A.; Wiedemann, U.A. Medium induced gluon radiation off massive quarks fills the dead cone. *Phys. Rev. D* **2004**, *69*, 114003. [CrossRef]
151. Das, S.K.; Scardina, F.; Plumari, S.; Greco, V. Heavy-flavor in-medium momentum evolution: Langevin versus Boltzmann approach. *Phys. Rev. C* **2014**, *90*, 044901. [CrossRef]
152. Li, S.; Wang, C.; Wan, R.; Liao, J. Probing the transport properties of Quark-Gluon Plasma via heavy-flavor Boltzmann and Langevin dynamics. *Phys. Rev. C* **2019**, *99*, 054909. [CrossRef]
153. Das, S.K.; Ruggieri, M.; Scardina, F.; Plumari, S.; Greco, V. Effect of pre-equilibrium phase on R_{AA} and v_2 of heavy quarks in heavy ion collisions. *J. Phys. G* **2017**, *44*, 095102. [CrossRef]
154. Lin, Z.W.; Di, G.T.; Ko, C.M. Charm meson scattering cross-sections by pion and rho meson. *Nucl. Phys. A* **2001**, *689*, 965–979. [CrossRef]
155. Cao, S.; Qin, G.Y.; Bass, S.A. Energy loss, hadronization and hadronic interactions of heavy flavors in relativistic heavy-ion collisions. *Phys. Rev. C* **2015**, *92*, 024907. [CrossRef]
156. Li, S.; Liao, J. Data-driven extraction of heavy quark diffusion in quark-gluon plasma. *Eur. Phys. J. C* **2020**, *80*, 671. [CrossRef]
157. Ding, H.T.; Karsch, F.; Mukherjee, S. Thermodynamics of strong-interaction matter from Lattice QCD. *Int. J. Mod. Phys. E* **2015**, *24*, 1530007. [CrossRef]
158. Banerjee, D.; Datta, S.; Gavai, R.; Majumdar, P. Heavy Quark Momentum Diffusion Coefficient from Lattice QCD. *Phys. Rev. D* **2012**, *85*, 014510. [CrossRef]
159. Kaczmarek, O. Continuum estimate of the heavy quark momentum diffusion coefficient κ. *Nucl. Phys. A* **2014**, *931*, 633–637. [CrossRef]
160. van Hees, H.; Rapp, R. Thermalization of heavy quarks in the quark-gluon plasma. *Phys. Rev. C* **2005**, *71*, 034907. [CrossRef]
161. Song, T.; Berrehrah, H.; Cabrera, D.; Torres-Rincon, J.M.; Tolos, L.; Cassing, W.; Bratkovskaya, E. Tomography of the Quark-Gluon-Plasma by Charm Quarks. *Phys. Rev. C* **2015**, *92*, 014910. [CrossRef]
162. Horowitz, W.A. Fluctuating heavy quark energy loss in a strongly coupled quark-gluon plasma. *Phys. Rev. D* **2015**, *91*, 085019. [CrossRef]
163. Xu, Y.; Bernhard, J.E.; Bass, S.A.; Nahrgang, M.; Cao, S. Data-driven analysis for the temperature and momentum dependence of the heavy-quark diffusion coefficient in relativistic heavy-ion collisions. *Phys. Rev. C* **2018**, *97*, 014907. [CrossRef]
164. Tolos, L.; Torres-Rincon, J.M. D-meson propagation in hot dense matter. *Phys. Rev. D* **2013**, *88*, 074019. [CrossRef]
165. Bialek, W.; Callan, C.G.; Strong, S.P. Field theories for learning probability distributions. *Phys. Rev. Lett.* **1996**, *77*, 4693–4697. [CrossRef]
166. Lemm, J.C. Bayesian field theory: Nonparametric approaches to density estimation, regression, classification, and inverse quantum problems. *arXiv* **1999**, arXiv:991205.
167. Xie, M.; Ke, W.; Zhang, H.; Wang, X.N. Global constraint on the jet transport coefficient from single hadron, dihadron and γ-hadron spectra in high-energy heavy-ion collisions. *arXiv* **2022**, arXiv:2208.14419.
168. Adam, J.; Adamczyk, L.; Adams, J.R.; Adkins, J.K.; Agakishiev, G.; Aggarwal, M.M.; Ahammed, Z.; Alekseev, I.; Anderson, D.M.; Aoyama, R.; et al. Evidence from d + Au measurements for final state suppression of high p(T) hadrons in Au+Au collisions at RHIC. *Phys. Rev. Lett.* **2003**, *91*, 072304. [CrossRef]
169. Miller, M.L.; Reygers, K.S.; Sanders, J.; Steinberg, P. Glauber modeling in high energy nuclear collisions. *Ann. Rev. Nucl. Part. Sci.* **2007**, *57*, 205–243. [CrossRef]
170. Adam, J.; Adamczyk, L.; Adams, J.R.; Adkins, J.K.; Agakishiev, G.; Aggarwal, M.M.; Ahammed, Z.; Alekseev, I.; Anderson, D.M.; Aoyama, R.; et al. Particle type dependence of azimuthal anisotropy and nuclear modification of particle production in Au + Au collisions at s(NN)**(1/2) = 200-GeV. *Phys. Rev. Lett.* **2004**, *92*, 052302. [CrossRef]
171. Adare, A.; Afanasiev, S.; Aidala, C.; Ajitanand, N.N.; Akiba, Y.; Al-Bataineh, H.; Alexander, J.; Al-Jamel, A.; Angerami, A.; Aoki, K.; et al. Nuclear modification factors of ϕ mesons in d+Au, Cu+Cu and Au+Au collisions at sqrt(S_NN) = 200 GeV. *Phys. Rev. C* **2011**, *83*, 024909. [CrossRef]
172. Adler, S.S.; Afanasiev, S.; Aidala, C.; Ajitanand, N.N.; Akiba, Y.; Alexander, J.; Amirikas, R.; Aphecetche, L.; Aronson, S.H.; Averbeck, R.; et al. Nuclear modification of electron spectra and implications for heavy quark energy loss in Au+Au collisions at s(NN)**(1/2) - 200-GeV. *Phys. Rev. Lett.* **2006**, *96*, 032301. [CrossRef] [PubMed]
173. Aad, G.; Abbott, B.; Abdallah, J.; Abdelalim, A.A.; Abdesselam, A.; Abdinov, O.; Abi, B.; Abolins, M.; Abramowicz, H.; Abreu, H.; et al. Measurements of the Nuclear Modification Factor for Jets in Pb+Pb Collisions at $\sqrt{s_{NN}} = 2.76$ TeV with the ATLAS Detector. *Phys. Rev. Lett.* **2015**, *114*, 072302. [CrossRef]
174. Aaboud, M.; Aad, G.; Abbott, B.; Abdallah, J.; Abdinov, O.; Abeloos, B.; Aben, R.; AbouZeid, O.; Abraham, N.; Abramowicz, H.; et al. Measurement of the nuclear modification factor for inclusive jets in Pb+Pb collisions at $\sqrt{s_{NN}} = 5.02$ TeV with the ATLAS detector. *Phys. Lett. B* **2019**, *790*, 108–128. [CrossRef]
175. Baier, R. Jet quenching. *Nucl. Phys. A* **2003**, *715*, 209–218. [CrossRef]
176. Burke, K.M.; Buzzatti, A.; Chang, N.; Gale, C.; Gyulassy, M.; Heinz, U.; Jeon, S.; Majumder, A.; Muller, B.; Qin, G.Y.; et al. Extracting the jet transport coefficient from jet quenching in high-energy heavy-ion collisions. *Phys. Rev. C* **2014**, *90*, 014909. [CrossRef]
177. Xie, M.; Wei, S.Y.; Qin, G.Y.; Zhang, H.Z. Extracting jet transport coefficient via single hadron and dihadron productions in high-energy heavy-ion collisions. *Eur. Phys. J. C* **2019**, *79*, 589. [CrossRef]

178. Ru, P.; Kang, Z.B.; Wang, E.; Xing, H.; Zhang, B.W. Global extraction of the jet transport coefficient in cold nuclear matter. *Phys. Rev. D* **2021**, *103*, L031901. [CrossRef]
179. Cao, S.; Chen, Y.; Coleman, J.; Mulligan, J.; Jacobs, P.M.; Soltz, R.A.; Angerami, A.; Arora, R.; Bass, S.A.; Cunqueiro, L.; et al. Determining the jet transport coefficient $\hat{q}$ from inclusive hadron suppression measurements using Bayesian parameter estimation. *Phys. Rev. C* **2021**, *104*, 024905. [CrossRef]
180. Huang, J.; Kang, Z.B.; Vitev, I. Inclusive b-jet production in heavy ion collisions at the LHC. *Phys. Lett. B* **2013**, *726*, 251. [CrossRef]
181. Frixione, S.; Webber, B.R. Matching NLO QCD computations and parton shower simulations. *JHEP* **2002**, *6*, 029. [CrossRef]
182. Ma, G.Y.; Dai, W.; Zhang, B.W.; Wang, E.K. NLO Productions of ω and K^0_S with a global extraction of the jet transport parameter in heavy-ion collisions. *Eur. Phys. J. C* **2019**, *79*, 518. [CrossRef]
183. Khachatryan, V.; Sirunyan, A.M.; Tumasyan, A.; Adam, W.; Asilar, E.; Bergauer, T.; Brandstetter, J.; Brondolin, E.; Dragicevic, M.; Erö, J.; et al. Measurement of inclusive jet cross-sections in pp and PbPb collisions at $\sqrt{s_{NN}} = 2.76$ TeV. *Phys. Rev. C* **2017**, *96*, 015202. [CrossRef]
184. Dokshitzer, Y.L.; Kharzeev, D.E. Heavy quark colorimetry of QCD matter. *Phys. Lett. B* **2001**, *519*, 199–206. [CrossRef]
185. Acharya, U.A.; Adare, A.; Aidala, C.; Ajitanand, N.N.; Akiba, Y.; Alfred, M.; Apadula, N.; Asano, H.; Azmoun, B.; Babintsev, V.; et al. Charm- and Bottom-Quark Production in Au+Au Collisions at $\sqrt{s_{NN}} = 200$ GeV. *arXiv* **2022**, arXiv:2203.17058.
186. Abdallah, M.S.; Aboona, B.E.; Adam, J.; Adamczyk, L.; Adams, J.R.; Adkins, J.K.; Aggarwal, I.; Aggarwal, M.M.; Ahammed, Z.; Anderson, D.M.; et al. Evidence of Mass Ordering of Charm and Bottom Quark Energy Loss in Au+Au Collisions at RHIC. *arXiv* **2021**, arXiv:2111.14615.
187. Ovanesyan, G.; Vitev, I. Medium-induced parton splitting kernels from Soft Collinear Effective Theory with Glauber gluons. *Phys. Lett. B* **2012**, *706*, 371. [CrossRef]
188. Sievert, M.D.; Vitev, I. Quark branching in QCD matter to any order in opacity beyond the soft gluon emission limit. *Phys. Rev. D* **2018**, *98*, 094010. [CrossRef]
189. Ke, W.; Wang, X.N.; Fan, W.; Bass, S.A. Study of heavy-flavor jets in a transport approach. *arXiv* **2020**, arXiv:2008.07622.
190. Wang, S.; Zhang, B.W.; Wang, E. Probing the mass effect of jet quenching by heavy-flavor jet in heavy-ion collisions. 2023. *In preparation.*
191. Wang, S.; Dai, W.; Zhang, B.W.; Wang, E. Z^0 Boson Associated b-jet Production in High-Energy Nuclear Collisions. *arXiv* **2020**, arXiv:2005.07018.
192. Kartvelishvili, V.; Kvatadze, R.; Shanidze, R. On Z and Z + jet production in heavy ion collisions. *Phys. Lett. B* **1995**, *356*, 589. [CrossRef]
193. Sirunyan, A.M.; Tumasyan, A.; Adam, W.; Ambrogi, F.; Asilar, E.; Bergauer, T.; Brandstetter, J.; Brondolin, E.; Dragicevic, M.; Erö, J.; et al. Study of jet quenching with isolated-photon+jet correlations in PbPb and pp collisions at $\sqrt{s_{NN}} = 5.02$ TeV. *Phys. Lett. B* **2018**, *785*, 14–39. [CrossRef]
194. Sirunyan, A.M.; Tumasyan, A.; Adam, W.; Ambrogi, F.; Asilar, E.; Bergauer, T.; Brandstetter, J.; Brondolin, E.; Dragicevic, M.; Erö, J.; et al. Study of Jet Quenching with $Z +$ jet Correlations in Pb-Pb and pp Collisions at $\sqrt{s_{NN}} = 5.02$ TeV. *Phys. Rev. Lett.* **2017**, *119*, 082301. [CrossRef] [PubMed]
195. Nason, P.; Dawson, S.; Ellis, R.K. The One Particle Inclusive Differential Cross-Section for Heavy Quark Production in Hadronic Collisions. *Nucl. Phys. B* **1989**, *327*, 49–92. [CrossRef]
196. Beenakker, W.; van Neerven, W.L.; Meng, R.; Schuler, G.A.; Smith, J. QCD corrections to heavy quark production in hadron hadron collisions. *Nucl. Phys. B* **1991**, *351*, 507–560. [CrossRef]
197. Mangano, M.L.; Nason, P.; Ridolfi, G. Heavy quark correlations in hadron collisions at next-to-leading order. *Nucl. Phys. B* **1992**, *373*, 295–345. [CrossRef]
198. Norrbin, E.; Sjostrand, T. Production and hadronization of heavy quarks. *Eur. Phys. J. C* **2000**, *17*, 137–161. [CrossRef]
199. Kang, Z.B.; Reiten, J.; Vitev, I.; Yoon, B. Light and heavy flavor dijets production and dijets mass modification in heavy ion collisions. *Phys. Rev. D* **2019**, *99*, 034006. [CrossRef]
200. Huang, J.; Kang, Z.B.; Vitev, I.; Xing, H. Photon-tagged and B-meson-tagged b-jet production at the LHC. *Phys. Lett. B* **2015**, *750*, 287. [CrossRef]
201. Gleisberg, T.; Hoeche, S. Comix, a new matrix element generator. *JHEP* **2008**, *12*, 039. [CrossRef]
202. Schumann, S.; Krauss, F. A Parton shower algorithm based on Catani-Seymour dipole factorisation. *JHEP* **2008**, *03*, 038. [CrossRef]
203. Mueller, A.H.; Wu, B.; Xiao, B.W.; Yuan, F. Probing Transverse Momentum Broadening in Heavy Ion Collisions. *Phys. Lett. B* **2016**, *763*, 208–212. [CrossRef]
204. Jia, J.; Wei, S.Y.; Xiao, B.W.; Yuan, F. Medium-Induced Transverse Momentum Broadening via Forward Dijet Correlations. *Phys. Rev. D* **2020**, *101*, 094008. [CrossRef]
205. Luo, T.; Cao, S.; He, Y.; Wang, X.N. Multiple jets and γ-jet correlation in high-energy heavy-ion collisions. *Phys. Lett. B* **2018**, *782*, 707–716. [CrossRef]
206. Zhang, S.L.; Luo, T.; Wang, X.N.; Zhang, B.W. Z+jet correlation with NLO-matched parton-shower and jet-medium interaction in high-energy nuclear collisions. *Phys. Rev. C* **2018**, *98*, 021901. [CrossRef]
207. Wang, S.; Dai, W.; Zhang, B.W.; Wang, E. The production of $b\bar{b}$ dijets in heavy-ion collisions at the LHC. *arXiv* **2018**, arXiv:1812.00391.

208. Adam, J.; Adamczyk, L.; Adams, J.R.; Adkins, J.K.; Agakishiev, G.; Aggarwal, M.M.; Ahammed, Z.; Alekseev, I.; Anderson, D.M.; Aoyama, R.; et al. Measurement of D^0-meson + hadron two-dimensional angular correlations in Au+Au collisions at $\sqrt{s_{NN}} = 200$ GeV. *Phys. Rev. C* **2020**, *102*, 014905. [CrossRef]
209. Wang, S.; Dai, W.; Yan, J.; Zhang, B.W.; Wang, E. Radial distribution of charm quarks in jets in high-energy heavy-ion collisions. *Nucl. Phys. A* **2021**, *1005*, 121787. [CrossRef]
210. Wang, S.; Dai, W.; Zhang, B.W.; Wang, E. Radial profile of heavy quarks in jets in high-energy nuclear collisions. *arXiv* **2020**, arXiv:2009.13959.
211. Wang, S.; Dai, W.; Zhang, B.W.; Wang, E. Radial profile of bottom quarks in jets in high-energy nuclear collisions. *Chin. Phys. C* **2021**, *45*, 064105. [CrossRef]
212. Renk, T. Biased showers: A common conceptual framework for the interpretation of high-P_T observables in heavy-ion collisions. *Phys. Rev. C* **2013**, *88*, 054902. [CrossRef]
213. Sjostrand, T.; Bengtsson, M. The Lund Monte Carlo for Jet Fragmentation and e+ e− Physics. Jetset Version 6.3: An Update. *Comput. Phys. Commun.* **1987**, *43*, 367. [CrossRef]
214. Webber, B.R. A QCD Model for Jet Fragmentation Including Soft Gluon Interference. *Nucl. Phys. B* **1984**, *238*, 492–528. [CrossRef]
215. Procura, M.; Stewart, I.W. Quark Fragmentation within an Identified Jet. *Phys. Rev. D* **2010**, *81*, 074009. [CrossRef]
216. Chatrchyan, S.; Khachatryan, V.; Sirunyan, A.M.; Tumasyan, A.; Adam, W.; Bergauer, T.; Dragicevic, M.; Erö, J.; Fabjan, C.; Friedl, M.; et al. Measurement of Jet Fragmentation in PbPb and pp Collisions at $\sqrt{s_{NN}} = 2.76$ TeV. *Phys. Rev. C* **2014**, *90*, 024908. [CrossRef]
217. Aad, G.; Abbott, B.; Abdallah, J.; Abdelalim, A.A.; Abdesselam, A.; Abdinov, O.; Abi, B.; Abolins, M.; Abramowicz, H.; Abreu, H.; et al. Measurement of inclusive jet charged-particle fragmentation functions in Pb+Pb collisions at $\sqrt{s_{NN}} = 2.76$ TeV with the ATLAS detector. *Phys. Lett. B* **2014**, *739*, 320–342. [CrossRef]
218. Aaboud, M.; Aad, G.; Abbott, B.; Abdallah, J.; Abdinov, O.; Abeloos, B.; Aben, R.; AbouZeid, O.; Abraham, N.; Abramowicz, H.; et al. Measurement of jet fragmentation in Pb+Pb and pp collisions at $\sqrt{s_{NN}} = 2.76$ TeV with the ATLAS detector at the LHC. *Eur. Phys. J. C* **2017**, *77*, 379. [CrossRef]
219. Aaboud, M.; Aad, G.; Abbott, B.; Abdallah, J.; Abdinov, O.; Abeloos, B.; Aben, R.; AbouZeid, O.; Abraham, N.; Abramowicz, H.; et al. Comparison of Fragmentation Functions for Jets Dominated by Light Quarks and Gluons from pp and Pb+Pb Collisions in ATLAS. *Phys. Rev. Lett.* **2019**, *123*, 042001.
220. ALICE Collaboration. Measurement of the production of charm jets tagged with D^0 mesons in pp collisions at $\sqrt{s} = 5.02$ and 13 TeV. *arXiv* **2022**, arXiv:2204.10167.
221. Li, Y.; Wang, S.; Zhang, B.W. Longitudinal momentum fraction of D^0 in jets in high-energy nuclear collisions. *arXiv* **2022**, arXiv:2209.00548.
222. Nason, P. A New method for combining NLO QCD with shower Monte Carlo algorithms. *JHEP* **2004**, *11*, 040. [CrossRef]
223. Alioli, S.; Nason, P.; Oleari, C.; Re, E. A general framework for implementing NLO calculations in shower Monte Carlo programs: The POWHEG BOX. *JHEP* **2010**, *06*, 043. [CrossRef]
224. Sjostrand, T.; Mrenna, S.; Skands, P.Z. A Brief Introduction to PYTHIA 8.1. *Comput. Phys. Commun.* **2008**, *178*, 852–867. [CrossRef]
225. Aad, G.; Abbott, B.; Abdallah, J.; Abdelalim, A.A.; Abdesselam, A.; Abdinov, O.; Abi, B.; Abolins, M.; Abramowicz, H.; Abreu, H.; et al. Measurement of the jet radius and transverse momentum dependence of inclusive jet suppression in lead-lead collisions at $\sqrt{s_{NN}} = 2.76$ TeV with the ATLAS detector. *Phys. Lett. B* **2013**, *719*, 220–241. [CrossRef]
226. Bossi, H. R-dependence of inclusive jet suppression and groomed jet splittings in heavy-ion collisions with ALICE. *arXiv* **2022**, arXiv:2208.14492.
227. Dokshitzer, Y.L.; Leder, G.D.; Moretti, S.; Webber, B.R. Better jet clustering algorithms. *JHEP* **1997**, *8*, 001. [CrossRef]
228. Larkoski, A.J.; Marzani, S.; Soyez, G.; Thaler, J. Soft Drop. *JHEP* **2014**, *5*, 146. [CrossRef]
229. Acharya, S.; Torales-Acosta, F.; Adamova, D.; Adolfsson, J.; Aggarwal, M.M.; Rinella, G.A.; Agnello, M.; Agrawal, N.; Ahammed, Z.; Ahn, S.U.; et al. Direct observation of the dead-cone effect in QCD. *Nature* **2022**, *605*, 440–446.
230. Dokshitzer, Y.L.; Khoze, V.A.; Troian, S.I. On specific QCD properties of heavy quark fragmentation ('dead cone'). *J. Phys. G* **1991**, *17*, 1602–1604. [CrossRef]
231. Cunqueiro, L.; Płoskoń, M. Searching for the dead cone effects with iterative declustering of heavy-flavor jets. *Phys. Rev. D* **2019**, *99*, 074027. [CrossRef]
232. Dai, W.; Li, M.Z.; Zhang, B.W.; Wang, E. Exposing the dead-cone effect of jet quenching in QCD medium. *arXiv* **2022**, arXiv:2205.14668.
233. Cunqueiro, L.; Napoletano, D.; Soto-Ontoso, A. Dead-cone searches in heavy-ion collisions using the jet tree. *arXiv* **2022**, arXiv:2211.11789.
234. Chang, N.B.; Tachibana, Y.; Qin, G.Y. Nuclear modification of jet shape for inclusive jets and γ-jets at the LHC energies. *Phys. Lett. B* **2020**, *801*, 135181. [CrossRef]
235. Goncalves, D.; Krauss, F.; Linten, R. Distinguishing b-quark and gluon jets with a tagged b-hadron. *Phys. Rev. D* **2016**, *93*, 053013. [CrossRef]
236. Ilten, P.; Rodd, N.L.; Thaler, J.; Williams, M. Disentangling Heavy Flavor at Colliders. *Phys. Rev. D* **2017**, *96*, 054019. [CrossRef]
237. ALICE Collaboration. Measurements of groomed-jet substructure of charm jets tagged by D^0 mesons in proton-proton collisions at $\sqrt{s} = 13$ TeV. *arXiv* **2022**, arXiv:2208.04857.

238. Zhang, Q.; Dai, W.; Zhang, B.W.; Wang, E. Substructures of heavy flavor jets in pp and Pb+Pb collisions at $\sqrt{s_{NN}}$ = 5.02 TeV. *In preparation*.
239. Li, H.T.; Vitev, I. Inverting the mass hierarchy of jet quenching effects with prompt *b*-jet substructure. *Phys. Lett. B* **2019**, *793*, 259. [CrossRef]
240. Sirunyan, A.M.; Tumasyan, A.; Adam, W.; Ambrogi, F.; Asilar, E.; Bergauer, T.; Brandstetter, J.; Brondolin, E.; Dragicevic, M.; Erö, J.; et al. Measurement of the Splitting Function in *pp* and Pb-Pb Collisions at $\sqrt{s_{NN}} = 5.02$ TeV. *Phys. Rev. Lett.* **2018**, *120*, 142302. [CrossRef]
241. Acharya, S.; Adamova, D.; Adler, A.; Adolfsson, J.; Aggarwal, M.M.; Aglieri Rinella, G.; Agnello, M.; Agrawal, N.; Ahammed, Z.; Ahmad, S.; et al. Exploration of jet substructure using iterative declustering in pp and Pb–Pb collisions at LHC energies. *Phys. Lett. B* **2020**, *802*, 135227. [CrossRef]

symmetry

MDPI

Review

Dilepton Program with Time-of-Flight Detector at the STAR Experiment

Xin Dong [1], Lijuan Ruan [2], Ming Shao [3], Yongjie Sun [3], Zebo Tang [3,*], Zhangbu Xu [2] and Wangmei Zha [3] and Yifei Zhang [3]

[1] Nuclear Science Division, Lawrence Berkeley National Laboratory, Berkeley, CA 94720, USA
[2] Physics Department, Brookhaven National Laboratory, Upton, NY 11973, USA
[3] State Key Laboratory of Particle Detection and Electronics, University of Science and Technology of China, Hefei 230026, China
* Correspondence: zbtang@ustc.edu.cn

Abstract: Pairs of lepton and antilepton (dilepton) in a continuous mass range are one of the most experimentally challenging and golden probes of the quark-gluon plasma (QGP) produced in heavy ion collisions because they do not strongly interact with the hot and dense medium, and reflect the properties of the medium at the time the dilepton is generated. The measurements of dileptons require lepton identification with high purity and high efficiency at large detector acceptance. STAR is one of two large experiments at the relativistic heavy ion collider with a primary goal of searching for the QGP and studying its properties. The STAR experiment launched a comprehensive dielectron (e^+e^-) program enabled by the time-of-flight (TOF) detector that had been fully installed in 2010. In this article, we review the decade-long *R&D*, the construction and performance of the STAR TOF detector, and dielectron measurements, including thermal dielectron production and dielectron production from the Breit–Wheeler process. Future perspectives are also discussed.

Keywords: quark gluon plasma; heavy ion collision; dilepton; chiral symmetry restoration; thermal radiation; strong magnetic field; Breit–Wheeler process

1. Introduction

Quark–gluon plasma (QGP) consists of fundamental building blocks of matter at the most extreme temperature and density conditions. QGP existed briefly during the first few microseconds after the Big Bang and can be recreated in ultrarelativistic heavy-ion collisions in which quarks and gluons are liberated from confinement within nucleons. The thermodynamic and hydrodynamic properties of the QGP, such as temperature and viscosity, have been under intense investigation by means of theoretical lattice quantum chromodynamic (QCD) computations [1,2], and in experiments utilizing the world's most powerful supercomputers and particle accelerators [3,4]. Thermal lepton pairs (e^+e^- and $\mu^+\mu^-$) from QGP radiation [5–7] are penetrating probes of the true temperature of the emitting source and are considered the holy grail of the study of QGP since they do not suffer from either final state interactions or the blue-shift effects from collective motion [8–11].

Despite the focus on nuclear physics, heavy-ion collisions have been understood as an opportunistic tool for studying quantum electrodynamics (QED) in unique regimes for nearly a century [12,13], and have attracted significant interest in the last 40 years [14–16], with generations of powerful accelerators. Some of the earliest measurements of electromagnetic processes in heavy-ion collisions were studied by the BEVALAC at LBNL [17], the AGS at BNL [18], and by the SPS at CERN [19]. In 2000, the relativistic heavy ion collider (RHIC) began operation and ushered in a new era of high-energy nuclear physics with ultrarelativistic Au beams, providing collisions with a center-of-mass energy per nucleon pair from $\sqrt{s_{NN}}$ = 200 GeV at top energy to 3.0 GeV in fixed-target mode. The payoff from this next-generation facility was almost immediate, with all four RHIC experiments reporting the observation of a new state of nuclear matter in 2004: the QGP [20–23]. In an influential workshop in 1990 titled "Can RHIC be used to test QED?" [24], the top priority was said to be "to understand the validity of the best available descriptions of e^+e^- pair production in peripheral heavy-ion collisions, especially for the domain where this process is known to be nonperturbative (multiple pair production)". It positively concluded that "A study of electromagnetic phenomena in extremely peripheral collisions of relativistic heavy ions can become a rich and exciting field that will complement studies of central collisions." However, none of the three (Landau–Lifschtz, Bethe–Heitler and Breit–Wheeler) processes was explicitly mentioned. Right from the start of RHIC operation, the two "large" experiments at RHIC (STAR and PHENIX) studied the electromagnetic production of e^+e^- from the electromagnetic fields of colliding ultrarelativistic Au nuclei [25,26]. Multiple experiments performed measurements of the total cross-section of electromagnetically produced l^+l^- from ultraperipheral heavy-ion collisions [26–28] or from exclusive $p+p$ [29] collisions. Significant progress through multiple discoveries was achieved in the physics of photon interactions. Experimental measurements and theoretical descriptions have been progressing from the initial discoveries toward quantitative and precise comparisons. Polarized photons were used and proposed as a tool to test and define the photon Wigner function [30–36], to probe the properties of the QGP [37–43], to measure nuclear charge and mass radii [37,44–46], to study gluon structure inside nuclei [47–49], and to investigate new quantum effects [45,48,50–53]. The subject has evolved significantly in the last decade, with new measurements and a new understanding of thermal radiation and strong QED fields.

With the installation of a time-of-flight detector (TOF) surrounding the STAR time projection chamber (TPC) starting in 2009–2010, the combination demonstrated effective electron particle identification with the necessary purity and large acceptance in a very cost-effective configuration. We dedicate this review article to the relevant STAR TOF detector at the RHIC facility, with special attention given to its unique capabilities and limitations with respect to measuring thermal dielectron radiation from QGP and the novel QED phenomena, especially the Breit–Wheeler process and vacuum birefringence.

2. Time-of-Flight Detector

The multigap resistive plate chamber (MRPC) was first developed by the ALICE TOF group in the 1990s [54] to efficiently identify copiously produced charged particles in relativistic heavy ion collisions. As a new type of gaseous detector with precision timing performance, the MRPC is suitable for applications requiring a large area, high granularity, and low cost.

The MRPC in China was first developed by the high-energy physics group led by Prof. Hongfang Chen at the University of Science and Technology of China (USTC) in 2000, when the STAR experiment at RHIC sought to build an MRPC-based TOF to improve the identification capability of charged particles. USTC was responsible for the *R&D* of MRPC to meet the requirements of STAR TOF.

The first MRPC prototype was very simple, with only one readout pad and a sensitive area of about 3×3 cm^2 and five 0.22 mm thick gas gaps (see the left panel of Figure 1), aiming to verify the feasibility of this technique. In late 2000, the performance of this prototype was studied in the T10 test beam facility at CERN. The prototype functioned well, with a time resolution of 70 ps and detection efficiency of >95% (right panel of Figure 1) [55,56]. Soon after this success, a batch of MRPC modules with different structures were developed and tested [57]. Figure 2 shows the structure of an MRPC prototype with a sensitive area of 20×6.3 cm^2 and six gas gaps. The sensitive area was divided into 12 regions by 12 readout pads with a size of 3.1×3.0 cm^2. The time resolution of this prototype reached 60 ps, and its detection efficiency exceeded 97% in the beam test with good uniformity [57]. An MRPC prototype with the same sensitive area, but with two stacks of five gas gaps, was also constructed and tested. Its time resolution reached 50 ps, and its detection efficiency exceeded 99%. Considering the limited available space for STAR TOF installation, the 6-gap MRPC structure was chosen for the STAR TOF. The readout pads were recombined into a size of 3.1×6.1 cm^2 to reduce the total number of readout channels [58].

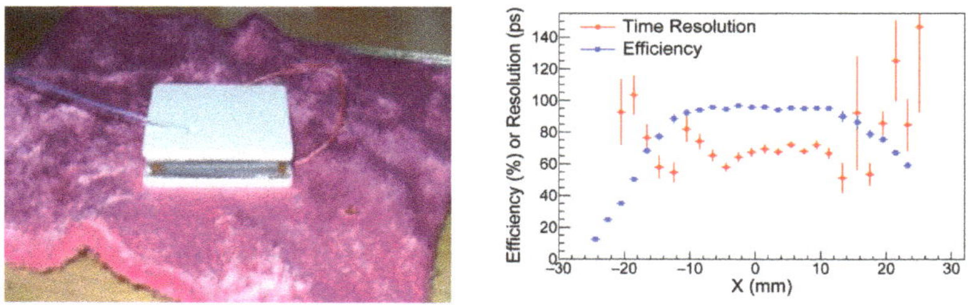

Figure 1. (**left**) The first MRPC module produced at USTC. (**right**) Detection efficiency and time resolution as a function of distance from the center of the readout pad [55].

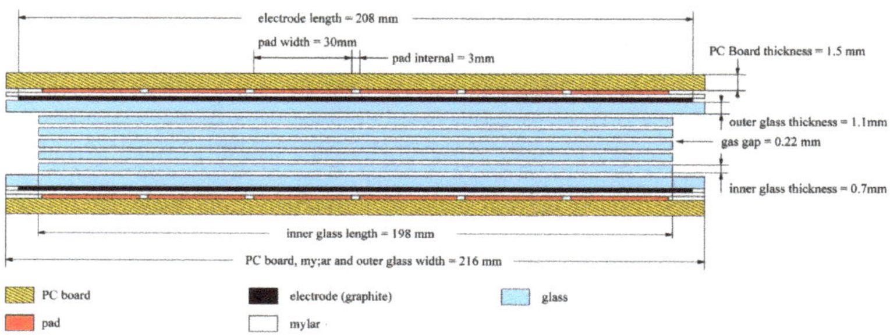

Figure 2. The structure of the MRPC prototype for STAR TOF.

In 2002, a TOF tray prototype (TOFr) with 28 MRPC modules was built by USTC and Rice university. A team led by Prof. Hongfang Chen tested the performance of the TOFr under various operating conditions at the AGS facility for 10 weeks prior to installation in the STAR experiment for the 2002–2003 physics run. The optimal parameters of high voltage, gas mixture, and threshold voltage for the operation of the prototype at STAR were determined from the beam test results [59]. TOFr was installed in STAR after the test at the AGS replacing one central trigger barrel module, and was part of the p+p and d+Au physics runs. In the STAR environment, the detection efficiency of the TOFr tray, averaged over all the channels, was around 95%, and the time resolution was 85 ps. The performance of the TOFr tray met the STAR physics goals and could significantly extend STAR particle identification capabilities [60]. With 2σ separation, protons/(pions + kaons) and kaons/pions were identified up to 3 and 1.6 GeV/c, respectively. Although the acceptance of the prototype was only 1/120 of the TOF system, several important physics results were achieved from the acquired data thanks to the extended particle identification capability [61,62]. With the success of the TOFr physics run, the STAR collaboration decided that all TOF MRPC modules were to be produced in China.

Thanks to the experiences of producing and operating the TOFr, and the extensive study of the MRPC technology [63–66], the production-process and quality-control requirements for MRPC module mass production for the STAR TOF were established. In 2006, funded by the National Natural Science Foundation of China, the Chinese Academy of Sciences, and the Ministry of Science and Technology, the Chinese STAR group launched the construction of the STAR TOF. By 2009, over 4000 MRPC modules were produced by Tsinghua University and USTC. Due to the strict quality management, the final yield was >95%, with excellent uniformity in performance [67,68]. The whole STAR barrel TOF consisted of 120 trays (each tray was analogous to the TOFr), 3840 MRPC modules, and

23,040 channels in total. It covered approximately the pseudo-rapidity interval $(-1, 1)$ and full azimuthal range. Operating with a 94.7% $C_2H_2F_4 + 5.3\%$ iso-Butane gas mixture, the MRPCs of the STAR barrel TOF showed a better intrinsic time resolution than 70 ps (the overall time resolution of STAR TOF also depended on the time jitter of the "start" time detector, which varied with different heavy ion collision systems). This time resolution and the corresponding particle identification capability were consistent with those achieved by TOFr and remained stable for more than 10 years of STAR operation.

After the success of the STAR TOF project, the USTC group continued to develop the MRPC technique and applied it to large facilities, such as a muon telescope detector and end-cap TOF for the STAR experiment, an end-cap TOF upgrade for the Beijing Spectrometer (BESIII), the TOF for the compressed baryonic matter (CBM) experiment, and the TOF for the CSR external-target experiment (CEE) [69].

3. Thermal Dileptons

Photons and dileptons (e^+e^- or $\mu^+\mu^-$) are emitted at various stages during the space-time evolution of the nuclear medium created in ultrarelativistic heavy-ion collisions. As penetrating electromagnetic probe dileptons do not suffer from strong interactions, they keep undistorted information of the sources from which the dileptons are coming. These sources are expected to contribute differently in lepton-pair invariant mass (M_{ll}), which are usually categorized into three regions:

(1) In the low-mass region (LMR), below ϕ mass ($M_{ll} < 1.1$ GeV/c^2), the main contributions are from light meson (π^0, η, ρ^0, ω, ϕ) decays or Dalitz decays. In particular, one can investigate hadronic in-medium properties via ρ^0 spectral modifications, which is sensitive to the mechanisms of chiral symmetry restoration in QCD matter [70]. Modified dilepton yields in LMR are expected to be related with the medium's lifetime and the transition from hadronic into partonic degrees of freedom [71].

(2) In the intermediate-mass region (IMR), namely, between ϕ and J/ψ mass ($M_{ll} \simeq 1.2$–3 GeV/c^2), the invariant-mass spectrum is continuum-like in both heavy flavor decays and nuclear matter emissions. This provides the chance to measure the direct radiation signals of QGP, which are expected to be a clean thermometer of the nuclear medium [10] by extracting the inverse slope of the mass spectra, which is unaffected by the blue shift of the expanding system.

(3) In the high-mass region (HMR), $M_{ll} \geq 3$ GeV/c^2, the main sources are heavy flavor/quarkonium decays and the Drell–Yan process.

Thermal dileptons from QGP radiation are of particular interest since they are sensitive to the properties of the produced medium. By investigating QGP radiation in different invariant-mass regions, one can probe different stages during the time evolution of the QGP phase.

The measurements of the above dileptons require large detector acceptance, large event statistics, and clean lepton identification. STAR collected a large sample of Au + Au collision events with the TOF detector. The TOF detector greatly extended the capability of the STAR experiment for particle identification. Combining the timing measurement from the TOF detector, and the momentum and ionization energy loss (dE/dx) measurements from the tracking detector, namely, the time projection chamber (TPC), one can cleanly identify low-momentum electrons with high efficiency and purity. Figure 3a shows the nσ_e (the n times of the standard deviation of measured dE/dx to the Bichsel prediction) as a function of particle momentum. The electron band largely overlapped with charged hadrons when using data from the TPC alone. With a TOF velocity close to the speed of light in vacuum, most of the hadrons were rejected. The remaining hadrons were a clean electron band and fast hadrons, as shown in Figure 3b.

The wide, uniform pseudorapidity and azimuthal acceptance of the STAR TPC and TOF detector allow for the precision reconstruction of dielectron pairs with the good systematic control of uncorrelated and correlated backgrounds. After the removal of these background contributions, dielectron invariant-mass spectra were measured with STAR at various collision energy levels. Decays from light mesons after a freeze-out (usually called a cocktail) can be removed by simulating the cocktail contributions. The remaining dielectron spectra are of particular interest and expected to contain radiation information from various stages of heavy-ion collisions before freeze-out.

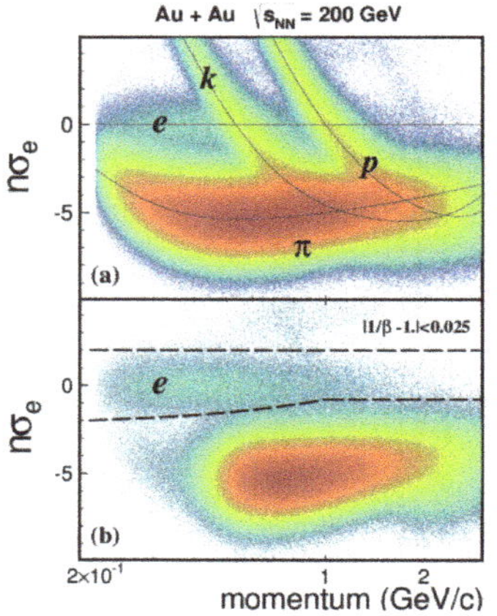

Figure 3. (**a**) nσ_e measured with the time projection chamber in Au + Au collisions at $\sqrt{s_{NN}}$ = 200 GeV. (**b**) nσ_e with a particle velocity cut using the TOF detector. The figure was taken from [72].

Figure 4 shows e^+e^- invariant-mass spectra after cocktail subtraction in Au + Au collisions at $\sqrt{s_{NN}}$ = 27, 39, 62.4, and 200 GeV [73,74]. The figure shows model calculations for total thermal radiation (solid lines), including both in-medium hadronic (dashed lines) and QGP (dotted lines) contributions. The model calculations provide a consistent description of the measured dielectron spectra across broad-energy and invariant-mass regions. In the LMR, the dominant hadronic radiation is mainly generated through ρ meson in-medium broadening through interactions with the hadronic medium, particularly baryons. The same model also provided a consistent description of the invariant-mass spectrum of the dimuon pairs measured in the NA60 experiment at SPS [75]. The ρ meson in-medium broadening was suggested to be evidence of the partial restoration of the chiral symmetry in the hot QCD medium [76]. The QGP radiation contribution was expected to take over at the IMR, where the precision of the current measurements is limited. The search and investigation of the QGP thermal radiation are future directions in the dilepton programs at both RHIC and LHC.

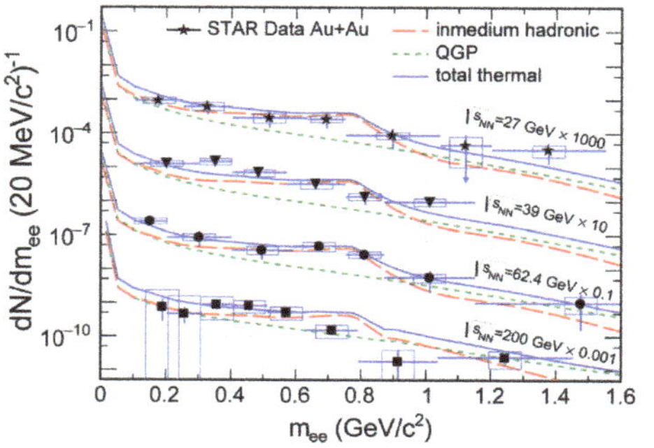

Figure 4. STAR measurements of dielectron invariant-mass spectra after cocktail removal in Au + Au collisions at $\sqrt{s_{NN}}$ = 27, 39, 62.4, and 200 GeV [73,74] in comparison with the model calculations of the total thermal radiation (blue solid lines), including in-medium hadronic (red dashed lines) and QGP (green dotted lines) contributions [10,71,77].

Figure 5 shows the collision energy dependence of the integrated dielectron yield, normalized to the charge pion dN/dy in the low-mass region of $0.3 < M_{ll} < 0.7$ GeV/c^2 measured from HADES [78], NA60 [75], and STAR [73,74]. Theoretical model calculations on dielectron yields (dashed blue line) and the fireball lifetime (solid red line) are also shown in the figure. The model provides a good description of the energy dependence, which exhibited a modest increase from the SPS to the top RHIC energy. This increase also shows the good tracking of the fireball lifetime across a broad collision energy range. The measurements from STAR shown here are from the Beam Energy Scan (BES) Phase I. The recorded BES-II data and analyses further extend the measurements from 19.6 GeV down to 7.7 GeV, offering new insights of the hot medium properties in the high-baryon-density region. Future experiments, such as NA60+, CBM and MPD, could further extend the measurements to low energy [79–81].

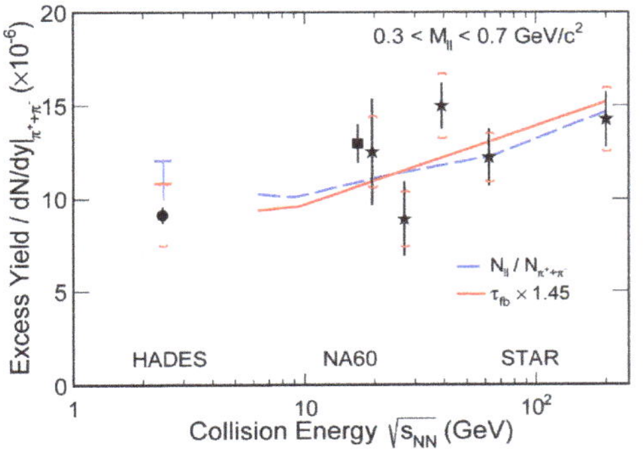

Figure 5. Collision energy dependence of the integrated dielectron yield normalized to the charged pion dN/dy in the low-mass region of $0.3 < M_{ll} < 0.7\,\text{GeV/c}^2$ measured from HADES [78], NA60 [75] and STAR [73,74] in comparison with the model calculations of dielectron yields (dashed blue line) and the fireball lifetime (solid red line) [77].

4. Dileptons from the Breit–Wheeler Process

When matter and antimatter [82] interact in a process called annihilation, they both disappear, and their mass is instantly converted into energy [83,84]. In 1934, Breit and Wheeler started the theoretical study of the simplest reverse process of the collision of two light quanta ($\gamma\gamma \to e^+e^-$) to create mass from energy [85]. The Breit–Wheeler study indicated that it is impossible to achieve this process via γ-ray collisions in existing Earth base experiments, which still holds true. Alternatively, they proposed observing the process via relativistic heavy-ion collisions in which the highly Lorentz-contracted electromagnetic field induced by fast-moving ions can be treated as a powerful source of photons for collisions. The treatment of quantizing the Lorentz-boosted Coulomb field was first proposed by Fermi [86], and further developed by Williams and Weizsacker [13], which is called the Williams–Weizsacker method or equivalent photon approximation (EPA).

Over the past few decades, dielectron production from the collision of light quanta was observed at various hadron and electron–positron collider experiments [19,25–27,29,87–92]. However, the existing experimental results lack precision and differential measurements to demonstrate themselves as the Breit–Wheeler process. The pioneering study of the Breit–Wheeler process at RHIC was conducted by STAR with Au + Au collision data from $\sqrt{s_{NN}}$ = 200 GeV taken in 2001 [25]. The data were collected with the trigger of ultraperipheral collisions (UPCs) in order to reject hadronic interactions. The specific energy loss (dE/dx) from the TPC only allows for the identification of electrons from pion backgrounds in a very limited kinematic range ($65 < p_T < 130$ MeV/c). In total, 52 candidate events were selected out of 800,000 triggered events in which the limited statistics prevents further claims from being made. For the data taken in 2010 with the fully installed TOF, the contamination background from hadron pairs could be largely reduced by using the double difference in the time of flight between the two measured tracks, and the expectation for dielectrons calculated using the measured momentum and path length from the TPC. Together with dE/dx information from the TPC, the optimized section criteria could ensure a 99% purity of dielectron candidates.

Measurements of the production rate of exclusive dielectron pairs from the data taken in 2010 [93] are shown in Figure 6. The measurements were performed in the fiducial phase space reported in the figure. Figure 6a shows the invariant-mass spectrum of exclusive dielectron pairs, which was smooth and featureless. The potential background contribution from exclusive vector meson photoproduction with the decay branch of dielectrons was estimated with STARLight [94] model and is shown as solid purple lines in the figure; the results were negligible in comparison with the experimental data. Figure 6b displays the differential $|\cos\theta'|$ distribution in which θ' is defined as the polar angle of the electron (positron) with respect to the beam in the electron–positron center-of-mass frame. The falloff of the distribution resulted from the detector acceptance. In comparison with isotropic emission (dashed line in the figure), the measured exclusive process exhibited enhancement toward a small polar angle. The differential cross-section, as a function of the pair transverse momentum ($P_\perp$), is shown in Figure 6c. The data characterize a clear peak in the production rate at very low transverse momentum. All these features were the consequence of the energy spectrum and quantum numbers of the two photons involved in the Breit–Wheeler process. Therefore, the Breit–Wheeler process was unambiguously identified for relativistic heavy-ion collisions in this measurement.

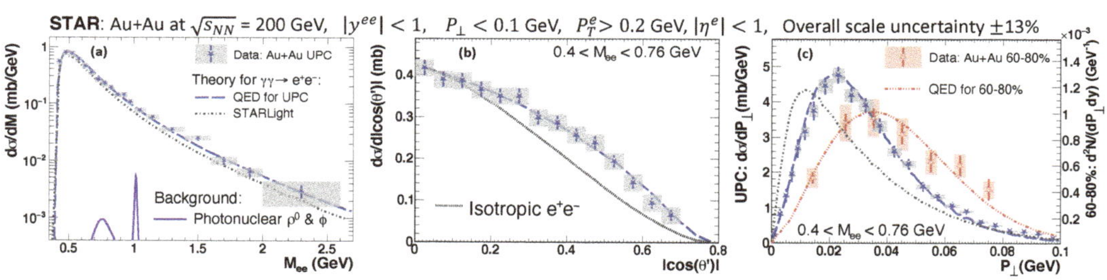

Figure 6. Differential cross-sections for exclusively dielectron pairs as functions of (**a**) invariant-mass M_{ee} (and potential vector-meson background from photoproduction), (**b**) polar angle distribution $|\cos\theta'|$, and (**c**) pair transverse momentum $P_\perp$. The figure was taken from [93].

Conventionally, the Breit–Wheeler process is only studied in UPCs to reject hadronic backgrounds. Furthermore, photon-induced processes are not generally considered to add up coherently for collisions with nuclear overlap. This begins with J/ψ measurements at a very low transverse momentum in Pb + Pb collisions from ALICE [28], in which significant excesses were observed in peripheral collisions. The abnormal excesses of J/ψ were confirmed via the STAR measurements [95], and well-described by the phenomenological model with a coherent photon-nucleus production mechanism [50,96–99]. Assuming that coherent photon nuclear production is responsible for the observed J/ψ excesses, the Breit–Wheeler process should also exist and contribute to dielectron production for collisions with nuclear overlap.

Utilizing the datasets collected in 2010, 2011, and 2012, STAR performed dielectron measurements at a low transverse momentum for hadronic heavy-ion collisions (non-UPCs) [38]. The electron identification criteria were the same as those described in Section 3. Figure 7 shows the transverse momentum distributions of dielectron pairs in different mass regions for 60–80% Au + Au and U + U collisions. As expected, a significant excess was observed and concentrated below $p_T \sim 0.15$ GeV/c with respect to the hadronic cocktail. The magnitude of the excess could be well-described with the model incorporated with the Breit–Wheeler process in hadronic heavy-ion collisions [100]. To further investigate the features of the Breit–Wheeler process in hadronic heavy-ion collisions, STAR measured the p_T^2 distributions of excess yields within acceptance in different mass regions for peripheral collisions, as shown in Figure 8. The aforementioned model calculations are also plotted in the figure. The theoretical results fell below the data points at large p_T^2 values, and overshot the data at low p_T^2. Such a discrepancy can be well-quantified by the corresponding $\sqrt{\langle p_T^2 \rangle}$ of p_T^2 distributions, as shown in Figure 8d. The disagreement between the data and the model calculations suggests the possible origins of the p_T broadening from the hot medium created in the nuclear overlap region. By introducing a magnetic field trapped in an electrically conducting QGP, the p_T broadening can be reasonably described. ATLAS collaboration also found a significant p_T broadening effect via the dimuon channel in comparison to those in UPCs [39], and quantified the effect via the acoplanarity of dimuon pairs. Alternatively, the broadening was explained by the electromagnetic scattering of leptons in the hot and dense medium [40]. All these descriptions of the broadening effect assumed that there was no impact-parameter dependence of the p_T distribution in the production process. Recent theoretical calculations [30,31] that recovered the impact parameter dependence demonstrated that the broadening was mainly from the initial light quantum collisions. This was further experimentally confirmed via CMS measurements [34] through exclusive dimuon production from photon–photon collisions in UPCs. Small tension still existed between data and theoretical calculations, which left room for possible effects from the hot medium.

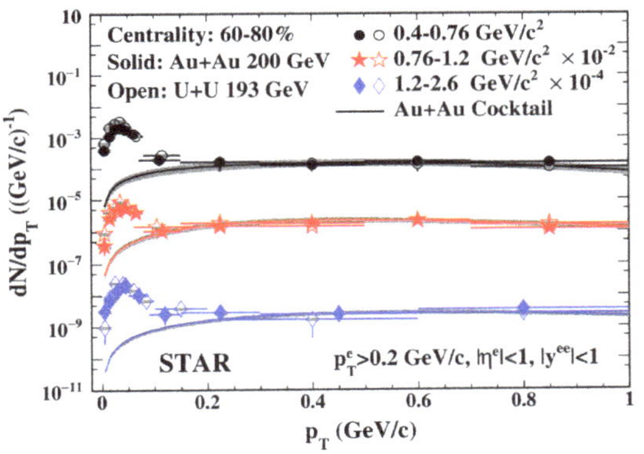

Figure 7. The transverse momentum distributions of dielectron pairs within STAR acceptance for different mass regions in peripheral Au + Au and U + U collisions compared to the hadronic cocktail. The figure was taken from [38].

Due to the extreme Lorentz contraction in relativistic heavy-ion collisions, the induced electromagnetic field was almost fully perpendicular to the direction of motion of the heavy nuclei, suggesting that the induced photons were fully linearly polarized in the transverse plane. Li et al. [101] proposed that the collisions of linearly polarized photons can lead to second- and fourth-order modulations in the azimuth (in the plane perpendicular to the beam direction). As shown in Figure 9, STAR performed the first measurement of the decay angular distribution ($\Delta\phi$) for the Breit–Wheeler process in relativistic heavy-ion collisions. Significant $\cos(4\Delta\phi)$ modulation was found, which could be described well via the numerical lowest-order QED calculation [102]. The observed angular modulation was closely related to the vacuum birefringence in which the vacuum was polarized by the extremely strong electromagnetic field in the absence of a medium [103]. This also inspired the discovery of $\cos(2\Delta\phi)$ modulation for vector meson photoproduction [45,48,51] in heavy-ion collisions.

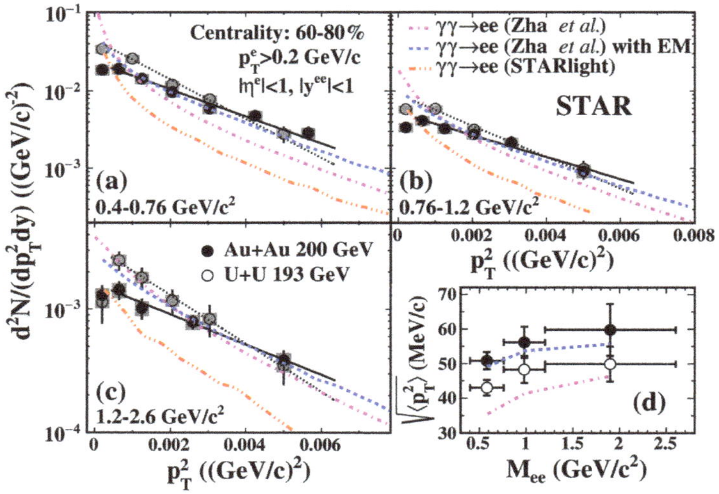

Figure 8. The p_T^2 distributions of excess yields within the STAR fiducial cuts in mass regions of (**a**) 0.40–0.76, (**b**) 0.76–1.2, and (**c**) 1.2–2.6 GeV/c^2 for peripheral Au + Au and U + U collisions. (**d**) The corresponding $\sqrt{\langle p_T^2 \rangle}$ of p_T^2 distributions. The figure was taken from [38].

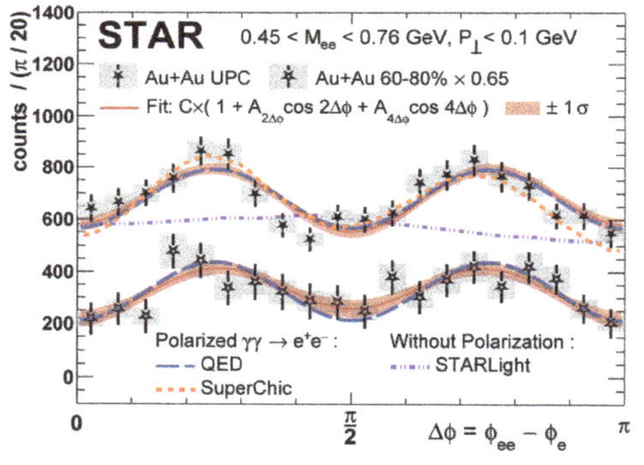

Figure 9. Decay angular distribution from UPCs and 60–80% central collisions with the theoretical calculations from QED [101], STARLight [94], and SUPERCHIC3 [104]. The figure was taken from [93].

5. Future Perspectives

The dilepton measurements at SPS with NA60 in the dimuon channel were performed with high statistics and with high data quality almost two decades ago. RHIC could provide important insights in the dielectron channel at the top energy and in the Beam Energy Scan Phase I (BES-I). The high-statistics data from BES-II and future data at the top energy level showed potential for providing the temperature measurements around the phase boundary and at the QGP phase. Future ALICE upgrades and experiments at high baryon density also promise an exciting dilepton program with LHC energies [79–81,105].

The future data-taking plan at RHIC and the LHC would provide an opportunity to perform the high-precision multidifferential analysis of these two photon-related physics [106,107]. The incoming RHIC run would allow for STAR to significantly improve the precision of the transverse momentum spectra and the angular modulation strength for the Breit–Wheeler process. The improved precision of transverse momentum spectra provides an additional powerful constraint to the proposed final-state broadening effects. The added precision in the angular modulation measurements is expected to experimentally demonstrate the impact parameter dependence of angular modulation strength, which is predicated by lowest-order QED calculations. Furthermore, precise angular modulation measurements could also serve as a sensitive probe of the hot medium, since the final state interactions would destroy the angular modulation resulting from the linear polarization of the initial photons. The much larger statistics provided by the future data-taking plan makes more differential measurements, such as angular modulation strength versus pair transverse momentum for dielectrons from the Breit–Wheeler process possible. The data-taking campaigns planned at the LHC would also allow for ALICE to measure the Breit–Wheeler process with much more precise measurements in a similar region of phase space as that measured at RHIC, but in collisions with a much larger Lorentz boost factor. Such measurements provide additional information on the photon kinematic distributions over a broad range of photon energy. Similarly, analyses of the $\gamma\gamma \to \mu^+\mu^-$ process in events with a hadronic overlap by CMS and ATLAS [107,108] from the future data-taking plan would shed new light on the presence (or lack) of medium-induced modifications. In much of the progress achieved in the last decade, TOF played an indispensable role. We anticipate that future progresses on this subject would continue to rely on the time-of-flight detector with its revolutionary technology advances.

Author Contributions: All authors contributed equally. All authors have read and agreed to the published version of the manuscript.

Funding: This review article received no external funding.

Data Availability Statement: The STAR published data are available at the STAR publication webpage: https://drupal.star.bnl.gov/STAR/publications (accessed on 27 December 2022) and HEPData: https://www.hepdata.net/ (accessed on 27 December 2022).

Conflicts of Interest: The authors declare no conflict of interest.

References

1. Bazavov, A.; Ding, H.T.; Hegde, P.; Kaczmarek, O.; Karsch, F.; Karthik, N.; Laermann, E.; Lahiri, A.; Larsen, R.; Li, S.T.; et al. Chiral crossover in QCD at zero and non-zero chemical potentials. *Phys. Lett. B* **2019**, *795*, 15–21.
2. Borsanyi, S.; Fodor, Z.; Guenther, J.N.; Kara, R.; Katz, S.D.; Parotto, P.; Pasztor, A.; Ratti, C.; Szabo, K.K. QCD Crossover at Finite Chemical Potential from Lattice Simulations. *Phys. Rev. Lett.* **2020**, *125*, 052001, [CrossRef] [PubMed]
3. Braun-Munzinger, P.; Koch, V.; Schäfer, T.; Stachel, J. Properties of hot and dense matter from relativistic heavy ion collisions. *Phys. Rept.* **2016**, *621*, 76–126. [CrossRef]
4. Busza, W.; Rajagopal, K.; van der Schee, W. Heavy Ion Collisions: The Big Picture, and the Big Questions. *Ann. Rev. Nucl. Part. Sci.* **2018**, *68*, 339–376. [CrossRef]
5. Shuryak, E.V. Quark-Gluon Plasma and Hadronic Production of Leptons, Photons and Psions. *Phys. Lett. B* **1978**, *78*, 150 [CrossRef]
6. McLerran, L.D.; Toimela, T. Photon and Dilepton Emission from the Quark - Gluon Plasma: Some General Considerations. *Phys. Rev. D* **1985**, *31*, 545 [CrossRef] [PubMed]
7. Shuryak, E.V.; Xiong, L. Dilepton and photon production in the 'hot glue' scenario. *Phys. Rev. Lett.* **1993**, *70*, 2241–2244. [CrossRef]
8. Adamczewski-Musch, J.; Arnold, O.; Behnke, C.; Belounnas, A.; Belyaev, A.; Berger-Chen, J.C.; Biernat, J.; Blanco, A.; Blume, C.; Böhmer, M.; et al. Probing dense baryon-rich matter with virtual photons. *Nat. Phys.* **2019**, *15*, 1040–1045 [CrossRef]
9. van Hees, H.; Rapp, R. Comprehensive interpretation of thermal dileptons at the SPS. *Phys. Rev. Lett.* **2006**, *97*, 102301, [CrossRef]
10. Rapp, R.; van Hees, H. Thermal Dileptons as Fireball Thermometer and Chronometer. *Phys. Lett. B* **2016**, *753*, 586–590. [CrossRef]
11. Shen, C.; Heinz, U.W.; Paquet, J.F.; Gale, C. Thermal photons as a quark-gluon plasma thermometer reexamined. *Phys. Rev. C* **2014**, *89*, 044910. [CrossRef]
12. Toll, J.S. The Dispersion Relation for Light and Its Application to Problems Involving Electron Pairs. Ph.D. Thesis, Princeton University, Princeton, NJ, USA, 1952.
13. Williams, E.J. Nature of the High Energy Particles of Penetrating Radiation and Status of Ionization and Radiation Formulae. *Phys. Rev.* **1934**, *45*, 729–730 [CrossRef]
14. Eichler, J. Theory of relativistic ion-atom collisions. *Phys. Rep.* **1990**, *193*, 165–277 [CrossRef]
15. Baur, G. Multiple electron-positron pair production in relativistic heavy-ion collisions: A strong-field effect. *Phys. Rev. A* **1990**, *42*, 5736–5738 [CrossRef]
16. Baltz, A.J.; Gelis, F.; McLerran, L.; Peshier, A. Coulomb corrections to e^+e^- production in ultra-relativistic nuclear collisions. *Nucl. Phys. A* **2001**, *695*, 395–429 [CrossRef]

17. Belkacem, A.; Gould, H.; Feinberg, B.; Bossingham, R.; Meyerhof, W.E. Capture, ionization, and pair-production processes in relativistic heavy-ion collisions in the 1-GeV/nucleon energy range. *Phys. Rev. A* **1997**, *56*, 2806–2818 [CrossRef]
18. Belkacem, A.; Claytor, N.; Dinneen, T.; Feinberg, B.; Gould, H. Electron capture from pair production by Au^{79+} at 10.8 GeV/nucleon. *Phys. Rev. A* **1998**, *58*, 1253–1255 [CrossRef]
19. Vane, C.R.; Datz, S.; Dittner, P.F.; Krause, H.F.; Bottcher, C.; Strayer, M.; Schuch, R.; Gao, H.; Hutton, R. Electron-positron pair production in Coulomb collisions of ultrarelativistic sulfur ions with fixed targets. *Phys. Rev. Lett.* **1992**, *69*, 1911–1914 [CrossRef]
20. Arsene, I.; Bearden, I.G.; Beavis, D.; Besliu, C.; Budick, B.; Bøggild, H.; Chasman, C.; Christensen, C.H.; Christiansen, P.; Cibor, J.; et al. Quark–gluon plasma and color glass condensate at RHIC? The perspective from the BRAHMS experiment. *Nucl. Phys. A* **2005**, *757*, 1–27. [CrossRef]
21. Back, B.B.; Baker, M.D.; Ballintijn, M.; Barton, D.S.; Becker, B.; Betts, R.R.; Bickley, A.A.; Bindel, R.; Budzanowski, A.; Busza, W.; et al. The PHOBOS perspective on discoveries at RHIC. *Nucl. Phys. A* **2005**, *757*, 28–101. [CrossRef]
22. Adams, J.; Aggarwal, M.M.; Ahammed, Z.; Amonett, J.; Anderson, B.D.; Arkhipkin, D.; Averichev, G.S.; Badyal, S.K.; Bai, Y.; Balewski, J.; et al. Experimental and theoretical challenges in the search for the quark–gluon plasma: The STAR Collaboration's critical assessment of the evidence from RHIC collisions. *Nucl. Phys. A* **2005**, *757*, 102–183. [CrossRef]
23. Adcox, K.; Adler, S.S.; Afanasiev, S.; Aidala, C.; Ajitan, N.N.; Akiba, Y.; Al-Jamel, A.; Alexer, J.; Amirikas, R.; Aoki, K.; et al. Formation of dense partonic matter in relativistic nucleus–nucleus collisions at RHIC: Experimental evaluation by the PHENIX Collaboration. *Nucl. Phys. A* **2005**, *757*, 184–283. [CrossRef]
24. Fatyga, M.; Rhoades-Brown, M.; Tannenbaum, M. *Can RHIC Be Used to Test QED?* Technical Report; Brookhaven National Lab.: Upton, NY, USA, 1990.
25. Adams, J.; Aggarwal, M.M.; Ahammed, Z.; Amonett, J.; Anderson, B.D.; Arkhipkin, D.; Averichev, G.S.; Bai, Y.; Balewski, J.; Barannikova, O.; et al. Production of e^+e^- pairs accompanied by nuclear dissociation in ultraperipheral heavy-ion collisions. *Phys. Rev. C* **2004**, *70*, 031902 [CrossRef]
26. Drachenberg, J.L.; Hagiwara, M.N.; Isenhower, D.; Isenhower, L.; Omiwade, O.O.; Smith, W.C.; Towell, R.S.; Singh, C.P.; Singh, V.; Tuli, S.K.; et al. Photoproduction of J/ψ and of high mass e^+e^- in ultra-peripheral Au+Au collisions at $\sqrt{s_{NN}} = 200$ GeV. *Phys. Lett. B* **2009**, *679*, 321–329. [CrossRef]
27. Abbas, E.; Abelev, B.; Adam, J.; Adamová, D.; Adare, A.M.; Aggarwal, M.M.; Aglieri Rinella, G.; Agnello, M.; Agocs, A.G.; Agostinelli, A.; et al. Charmonium and e^+e^- pair photoproduction at mid-rapidity in ultra-peripheral Pb-Pb collisions at $\sqrt{s_{NN}} = 2.76$ TeV. *Eur. Phys. J. C* **2013**, *73*, 2617. [CrossRef] [PubMed]
28. Adam, J.; Adamová, D.; Aggarwal, M.M.; Rinella, G.A.; Agnello, M.; Agrawal, N.; Ahammed, Z.; Ahn, S.U.; Aiola, S.; Akindinov, A.; et al. Measurement of an excess in the yield of J/ψ at very low p_T in Pb-Pb collisions at $\sqrt{s_{NN}} = 2.76$ TeV. *Phys. Rev. Lett.* **2016**, *116*, 222301. [CrossRef]
29. Barreiro Alonso, F.; Aad, G.; Arnal, V.; Cantero, J.; De la Torre, H.; Del Peso, J.; Glasman, C.; Llorente, Merino, J.; Terrón, J., ATLAS Collaboration. Measurement of exclusive $\gamma\gamma \rightarrow l^+l^-$ production in proton–proton collisions at $\sqrt{s} = 7$ TeV with the ATLAS detector. *Phys. Lett. B* **2015**, *749*, 242–261. [CrossRef]
30. Zha, W.; Brandenburg, J.D.; Tang, Z.; Xu, Z. Initial transverse-momentum broadening of Breit-Wheeler process in relativistic heavy-ion collisions. *Phys. Lett. B* **2020**, *800*, 135089. [CrossRef]
31. Wang, R.j.; Pu, S.; Wang, Q. Lepton pair production in ultraperipheral collisions. *Phys. Rev. D* **2021**, *104*, 056011. [CrossRef]
32. Klein, S.; Mueller, A.H.; Xiao, B.W.; Yuan, F. Lepton Pair Production Through Two Photon Process in Heavy Ion Collisions. *Phys. Rev. D* **2020**, *102*, 094013. [CrossRef]
33. Kłusek-Gawenda, M.; Schäfer, W.; Szczurek, A. Centrality dependence of dilepton production via $\gamma\gamma$ processes from Wigner distributions of photons in nuclei. *Phys. Lett. B* **2021**, *814*, 136114. [CrossRef]
34. Sirunyan, A.M.; Tumasyan, A.; Adam, W.; Bergauer, T.; Dragicevic, M.; Erö, J.; Del Valle, A.E.; Fruehwirth, R.; Jeitler, M.; Krammer, N.; et al. Observation of Forward Neutron Multiplicity Dependence of Dimuon Acoplanarity in Ultraperipheral Pb-Pb Collisions at $\sqrt{s_{NN}} = 5.02$ TeV. *Phys. Rev. Lett.* **2021**, *127*, 122001. [CrossRef] [PubMed]
35. Sun, Z.H.; Zheng, D.X.; Zhou, J.; Zhou, Y.J. Studying Coulomb correction at EIC and EicC. *Phys. Lett. B* **2020**, *808*, 135679. [CrossRef]
36. Zha, W.; Tang, Z. Discovery of higher-order quantum electrodynamics effect for the vacuum pair production. *J. High Energy Phys.* **2021**, *2021*, 83. [CrossRef]
37. Brandenburg, J.D.; Zha, W.; Xu, Z. Mapping the electromagnetic fields of heavy-ion collisions with the Breit-Wheeler process. *Eur. Phys. J. A* **2021**, *57*, 299. [CrossRef]
38. STAR Collaboration; Adam, J. Low-p_T e^+e^- Pair Production in Au+Au Collisions at $\sqrt{s_{NN}} = 200$ GeV and U+U Collisions at $\sqrt{s_{NN}} = 193$ GeV at STAR. *Phys. Rev. Lett.* **2018**, *121*, 132301. [CrossRef]
39. Aaboud, M.; Aad, G.; Abbott, B.; Abeloos, B.; Abhayasinghe, D.K.; Abidi, S.H.; AbouZeid, O.S.; Abraham, N.L.; Abramowicz, H.; Abreu, H.; et al. Observation of Centrality-Dependent Acoplanarity for Muon Pairs Produced via Two-Photon Scattering in Pb + Pb Collisions at $\sqrt{s_{NN}} = 5.02$ TeV with the ATLAS Detector. *Phys. Rev. Lett.* **2018**, *121*, 212301. [CrossRef]
40. Klein, S.; Mueller, A.H.; Xiao, B.W.; Yuan, F. Acoplanarity of a Lepton Pair to Probe the Electromagnetic Property of Quark Matter. *Phys. Rev. Lett.* **2019**, *122*, 132301. [CrossRef]
41. Wang, Z.; Zhao, J.; Greiner, C.; Xu, Z.; Zhuang, P. Incomplete electromagnetic response of hot QCD matter. *Phys. Rev. C* **2022**, *105*, L041901. [CrossRef]

42. An X.; Bluhm, M.; Du, L.; Dunne, G.V.; Elfner, H.; Gale, C.; Grefa, J.; Heinz, U.; Huang, A.; Karthein, J.M.; et al. The BEST framework for the search for the QCD critical point and the chiral magnetic effect. *Nucl. Phys. A* **2022**, *1017*, 122343. [CrossRef]
43. Kłusek-Gawenda, M.; Rapp, R.; Schäfer, W.; Szczurek, A. Dilepton Radiation in Heavy-Ion Collisions at Small Transverse Momentum. *Phys. Lett. B* **2019**, *790*, 339–344. [CrossRef]
44. Wang, X.; Brandenburg, J.D.; Ruan, L.; Shao, F.; Xu, Z.; Yang, C.; Zha, W. Energy Dependence of the Breit-Wheeler process in Heavy-Ion Collisions and its Application to Nuclear Charge Radius Measurements. *arXiv* **2022**, arXiv:2207.05595.
45. Abdallah, M.S.; Aboona, B.E.; Adam, J.; Adamczyk, L.; Adams, J.R.; Adkins, J.K.; Agakishiev, G.; Aggarwal, I.; Aggarwal, M.M.; Ahammed, Z.; et al. Tomography of Ultra-relativistic Nuclei with Polarized Photon-gluon Collisions. *arXiv* **2022**, arXiv:2204.01625.
46. Budker, D.; Berengut, J.C.; Flambaum, V.V.; Gorchtein, M.; Jin, J.; Karbstein, F.; Krasny, M.W.; Litvinov, Y.A.; Pálffy, A.; Pascalutsa, V.; et al. Expanding Nuclear Physics Horizons with the Gamma Factory. *Ann. Phys.* **2021**, *534*, 2100284. [CrossRef]
47. Hatta, Y.; Xiao, B.W.; Yuan, F.; Zhou, J. Azimuthal angular asymmetry of soft gluon radiation in jet production. *Phys. Rev. D* **2021**, *104*, 054037. [CrossRef]
48. Xing, H.; Zhang, C.; Zhou, J.; Zhou, Y.J. The cos 2ϕ azimuthal asymmetry in ρ^0 meson production in ultraperipheral heavy ion collisions. *JHEP* **2020**, *10*, 064. [CrossRef]
49. Bor, J.; Boer, D. TMD Evolution Study of the cos 2ϕ Azimuthal Asymmetry in Unpolarized J/ψ Production at EIC. *arXiv* **2022**, arXiv:hep-ph/2204.01527.
50. Zha, W.; Ruan, L.; Tang, Z.; Xu, Z.; Yang, S. Double-slit experiment at fermi scale: Coherent photoproduction in heavy-ion collisions. *Phys. Rev. C* **2019**, *99*, 061901. [CrossRef]
51. Zha, W.; Brandenburg, J.D.; Ruan, L.; Tang, Z.; Xu, Z. Exploring the double-slit interference with linearly polarized photons. *Phys. Rev. D* **2021**, *103*, 033007. [CrossRef]
52. Dyndal, M.; Klusek-Gawenda, M.; Schott, M.; Szczurek, A. Anomalous electromagnetic moments of τ lepton in $\gamma\gamma \to \tau^+\tau^-$ reaction in Pb+Pb collisions at the LHC. *Phys. Lett. B* **2020**, *809*, 135682. [CrossRef]
53. Xu, I.; Lewis, N.; Wang, X.; Brandenburg, J.D.; Ruan, L. Search for Dark Photons in $\gamma\gamma \to e^+e^-$ at RHIC. *arXiv* **2022**, arXiv:2211.02132.
54. Cerron Zeballos, E.; Crotty, I.; Hatzifotiadou, D.; Lamas Valverde, J.; Neupane, S.; Williams, M.C.S.; Zichichi, A. A New type of resistive plate chamber: The Multigap RPC. *Nucl. Instrum. Meth. A* **1996**, *374*, 132–136. [CrossRef]
55. Li, C.; Wu, J.; Wang, X.; Chen, H.; Xu, Z.; Shao, M.; Ye, S.; Ruan, L.; Huang, S. A high time resolution multi-gap resistive plate chamber. *Nucl. Sci. Tech.* **2002**, *13*, 6–10.
56. Li, C.; Wu, J.; Chen, H.; Wang, X.; Xu, Z.; Wang, Z.; Shao, M.; Huang, S.; Ruan, L. A prototype of the high time resolution MRPC. *Chin. Phys. C* **2001**, *25*, 933–935.
57. Shao, M.; Ruan, L.J.; Chen, H.F.; Wu, J.; Li, C.; Xu, Z.Z.; Wang, X.L.; Huang, S.L.; Wang, Z.M.; Zhang, Z.P. Beam test results of two kinds of multi-gap resistive plate. *Nucl. Instrum. Meth. A* **2002**, *492*, 344–350. [CrossRef]
58. Wu, J.; Bonner, B.; Chen, H.F.; Dong, X.; Eppley, G.; Geurts, F.; Huang, S.L.; Li, C.; Llope, W.J.; Nussbaum, T.; et al. The performance of the TOFr tray in STAR. *Nucl. Instrum. Meth. A* **2005**, *538*, 243–248. [CrossRef]
59. Geurts, F.; Shao, M.; Bonner, B.; Chen, H.; Dong, X.; Eppley, G.; Huang, S.; Li, C.; Li, J.; Llope, W.; et al. Performance of the prototype MRPC detector for STAR. *Nucl. Instrum. Meth. A* **2004**, *533*, 60–64. [CrossRef]
60. Shao, M.; Barannikova, O.Y.; Dong, X.; Fisyak, Y.; Ruan, L.; Sorensen, P.; Xu, Z. Extensive particle identification with TPC and TOF at the STAR experiment. *Nucl. Instrum. Meth. A* **2006**, *558*, 419–429. [CrossRef]
61. Adams, J.; Aggarwal, M.M.; Ahammed, Z.; Amonett, J.; Anderson, B.D.; Arkhipkin, D.; Averichev, G.S.; Badyal, S.K.; Bai, Y.; Balewski, J.; et al. Pion, kaon, proton and anti-proton transverse momentum distributions from $p+p$ and d + Au collisions at $\sqrt{s_{NN}} = 200$ GeV. *Phys. Lett. B* **2005**, *616*, 8–16. [CrossRef]
62. Adams, J.; Aggarwal, M.M.; Ahammed, Z.; Amonett, J.; Anderson, B.D.; Arkhipkin, D.; Averichev, G.S.; Badyal, S.K.; Bai, Y.; Balewski, J.; et al. Open charm yields in d + Au collisions at $\sqrt{s_{NN}} = 200$ GeV. *Phys. Rev. Lett.* **2005**, *94*, 062301. [CrossRef]
63. Ruan, L.; Shao, M.; Chen, H.; Li, C.; Wang, X.; Wu, J.; Xu, Z.; Huang, S. A Monte Carlo Simulation of Multi-gap Resistive Plate Chamber and Comparison with Experimental Results. *Chin. Phys. C* **2003**, *27*, 712–715.
64. Zhao, Y.E.; Wang, X.L.; Liu, H.D.; Chen, H.F.; Li, C.; Wu, J.; Xu, Z.Z.; Shao, M.; Zeng, H.; Zhou, Y.; et al. Effect of temperature on the multi-gap resistive plate chamber operation. *Nucl. Instrum. Meth. A* **2005**, *547*, 334–341. [CrossRef]
65. Shao, M.; Zhao, Y.; Li, C.; Chen, H.F.; Wang, X.L.; Wu, J.; Sun, Y.J.; Ruan, L.J. Simulation study on the operation of a multi-gap resistive plate chamber. *Measur. Sci. Tech.* **2006**, *17*, 123–127. [CrossRef]
66. Shao, M.; Dong, X.; Tang, Z.; Xu, Y.; Huang, M.; Li, C.; Chen, H.F.; Lu, Y.; Zhang, Y. Upgrade of the calibration procedure for a STAR time-of-flight detector with new electronics. *Measur. Sci. Tech.* **2009**, *20*, 025102. [CrossRef]
67. Zou, T.; Wang, X.; Shao, M.; Sun, Y.; Zhao, Y.E.; Tang, H.; Ming, Y.; Guo, J.; Zhang, Y.; Li, C.; et al. Quality control of MRPC mass production for STAR TOF. *Nucl. Instrum. Meth. A* **2009**, *605*, 282–292. [CrossRef]
68. Llope, W.J. Multigap RPCs in the STAR experiment at RHIC. *Nucl. Instrum. Meth. A* **2012**, *661*, S110–S113. [CrossRef]
69. Shao, M.; Sun, Y.; Tang, Z.; Zhang, Y.; Zhang, Y.; Zhang, Z.; Zhao, L. Technology Development for Nuclear Physics at USTC. *Nucl. Phys. News* **2019**, *29*, 5–11. [CrossRef]
70. Hohler, P.M.; Rapp, R. Is ρ-Meson Melting Compatible with Chiral Restoration? *Phys. Lett. B* **2014**, *731*, 103–109. [CrossRef]

71. Rapp, R.; Wambach, J. Low mass dileptons at the CERN SPS: Evidence for chiral restoration? *Eur. Phys. J. A* **1999**, *6*, 415–420. [CrossRef]
72. Adamczyk, L.; Adkins, J.K.; Agakishiev, G.; Aggarwal, M.M.; Ahammed, Z.; Alekseev, I.; Alford, J.; Aparin, A.; Arkhipkin, D.; Aschenauer, E.C.; et al. Measurements of Dielectron Production in Au+Au Collisions at $\sqrt{s_{\mathrm{NN}}}$ = 200 GeV from the STAR Experiment. *Phys. Rev. C* **2015**, *92*, 024912. [CrossRef]
73. Adamczyk, L.; Adkins, J.K.; Agakishiev, G.; Aggarwal, M.M.; Ahammed, Z.; Alekseev, I.; Alford, J.; Aparin, A.; Arkhipkin, D.; Aschenauer, E.C.; et al. Energy dependence of acceptance-corrected dielectron excess mass spectrum at mid-rapidity in Au+Au collisions at $\sqrt{s_{\mathrm{NN}}} = 19.6$ and 200 GeV. *Phys. Lett. B* **2015**, *750*, 64–71. [CrossRef]
74. Adam, J.; Adamczyk, L.; Adams, J.R.; Adkins, J.K.; Agakishiev, G.; Aggarwal, M.; Ahammed, Z.; Alekseev, I.; Anderson, D.M.; Aoyama, R.; et al. Measurements of Dielectron Production in Au+Au Collisions at $\sqrt{s_{\mathrm{NN}}}$= 27, 39, and 62.4 GeV from the STAR Experiment. *arXiv* **2018**, arXiv:nucl-ex/1810.10159.
75. NA60 Collaboration; Arnaldi, R.; Banicz, K.; Borer, K.; Castor, J.; Chaur, B.; Chen, W.; Cicalò, C.; Colla, A.; Cortese, P.; et al. NA60 results on thermal dimuons. *Eur. Phys. J. C* **2009**, *61*, 711–720. [CrossRef]
76. HotQCD Collaboration; Bazavov, A.; Bhattacharya, T.; Buchoff, M.; Cheng, M.; Christ, N.; Ding, H.; Gupta, R.; Hegde, P.; Jung, C.; et al. The chiral transition and $U(1)_A$ symmetry restoration from lattice QCD using Domain Wall Fermions. *Phys. Rev. D* **2012**, *86*, 094503. [CrossRef]
77. van Hees, H.; Rapp, R. Dilepton Radiation at the CERN Super Proton Synchrotron. *Nucl. Phys. A* **2008**, *806*, 339–387. [CrossRef]
78. Agakishiev, G.; Balanda, A.; Belver, D.; Belyaev, A.; Blanco, A.; Böhmer, M.; Boyard, J.L.; Cabanelas, P.; Castro, E.; Chernenko, S.; et al. Dielectron production in Ar+KCl collisions at 1.76A GeV. *Phys. Rev. C* **2011**, *84*, 014902. [CrossRef]
79. Ahdida, C.; Alocco, G.; Antinori, F.; Arba, M.; Aresti, M.; Arnaldi, R.; Roldan, A.B.; Beole, S.; Beraudo, A.; Bernhard, J.; et al. Letter of Intent: The NA60+ experiment. *arXiv* **2022**, arXiv:2212.14452.
80. Abgaryan, V.; Acevedo Kado, R.; Afanasyev, S.V.; Agakishiev, G.N.; Alpatov, E.; Altsybeev, G.; Alvarado Hernández, M.; Andreeva, S.V.; Andreeva, T.V.; Andronov, E.V.; et al. Status and initial physics performance studies of the MPD experiment at NICA. *Eur. Phys. J. A* **2022**, *58*, 140. [CrossRef]
81. Ablyazimov, T.; Abuhoza, A.; Adak, R.P.; Adamczyk, M.; Agarwal, K.; Aggarwal, M.M.; Ahammed, Z.; Ahmad, F.; Ahmad, N.; Ahmad, S.; et al. Challenges in QCD matter physics –The scientific programme of the Compressed Baryonic Matter experiment at FAIR. *Eur. Phys. J. A* **2017**, *53*, 60. [CrossRef]
82. Anderson, C.D. The Apparent Existence of Easily Deflectable Positives. *Science* **1932**, *76*, 238–239. [CrossRef]
83. Chao, C.Y. Scattering of Hard γ-Rays. *Phys. Rev.* **1930**, *36*, 1519–1522. [CrossRef]
84. Klemperer, O. On the annihilation radiation of the positron. *Math. Proc. Camb. Philos. Soc.* **1934**, *30*, 347–354. [CrossRef]
85. Breit, G.; Wheeler, J.A. Collision of two light quanta. *Phys. Rev.* **1934**, *46*, 1087–1091. [CrossRef]
86. Fermi, E. On the Theory of the impact between atoms and electrically charged particles. *Z. Phys.* **1924**, *29*, 315–327. [CrossRef]
87. Bauer, R.; Breskin, A.; Chechik, R.; Drees, A.; Faschingbauer, U.; Fischer, P.; Fraenkel, Z.; Fuchs, C.; Gatti, E.; Gläß, J.; et al. Measurement of electromagnetically produced e^+e^- pairs in distant S-Pt collisions. *Phys. Lett. B* **1994**, *332*, 471–476. [CrossRef]
88. Acciarri, M.; Adriani, O.; Aguilar-Benitez, M.; Ahlen, S.; Alcaraz, J.; Alemanni, G.; Allaby, J.; Aloisio, A.; Alverson, G.; Alviggi, M.G.; et al. Production of e, μ and τ pairs in untagged two photon collisions at LEP. *Phys. Lett. B* **1997**, *407*, 341–350. [CrossRef]
89. Adeva, B.; Anderhub, H.; Ansari, S.; Becker, U.; Becker-Szendy, R.; Berdugo, J.; Boehm, A.; Bourquin, M.; Branson, J.G.; Burger, J.D.; et al. Electroweak studies in e^+e^- collisions: $12 < \sqrt{s} < 46.78$ GeV. *Phys. Rev. D* **1988**, *38*, 2665–2678. [CrossRef]
90. The OPAL Collaboration; Abbiendi, G. Total hadronic cross-section of photon-photon interactions at LEP. *Eur. Phys. J. C* **2000**, *14*, 199–212. [CrossRef]
91. The CMS collaboration.; Chatrchyan, S. Search for exclusive or semi-exclusive $\gamma\gamma$ production and observation of exclusive and semi-exclusive e^+e^- production in pp collisions at $\sqrt{s} = 7$ TeV. *J. High Energ. Phys.* **2012**, *2012*, 80.
92. The CMS Collaboration; TOTEM Collaboration; Sirunyan, A.M.; Tumasyan, A.; Adam, W.; Ambrogi, F.; Asilar, E.; Bergauer, T.; Brstetter, J.; Brondolin, E.; et al. Observation of proton-tagged, central (semi)exclusive production of high-mass lepton pairs in pp collisions at 13 TeV with the CMS-TOTEM precision proton spectrometer. *J. High Energ. Phys.* **2018**, *2018*, 153.
93. Abdallah, M.S.; Aboona, B.E.; Adam, J.; Adamczyk, L.; Adams, J.R.; Adkins, J.K.; Agakishiev, G.; Aggarwal, I.; Aggarwal, M.M.; Ahammed, Z.; et al. Measurement of e^+e^- Momentum and Angular Distributions from Linearly Polarized Photon Collisions. *Phys. Rev. Lett.* **2021**, *127*, 052302. [CrossRef]
94. Klein, S.R.; Nystrand, J.; Seger, J.; Gorbunov, Y.; Butterworth, J. STARlight: A Monte Carlo simulation program for ultra-peripheral collisions of relativistic ions. *Comput. Phys. Commun.* **2017**, *212*, 258–268. [CrossRef]
95. Adam, J.; Adamczyk, L.; Adams, J.R.; Adkins, J.K.; Agakishiev, G.; Aggarwal, M.M.; Ahammed, Z.; Alekseev, I.; Anderson, D.M.; Aoyama, R.; et al. Observation of excess J/ψ yield at very low transverse momenta in Au+Au collisions at $\sqrt{s_{\mathrm{NN}}} = 200$ GeV and U+U collisions at $\sqrt{s_{\mathrm{NN}}} = 193$ GeV. *Phys. Rev. Lett.* **2019**, *123*, 132302. [CrossRef] [PubMed]
96. Zha, W.; Klein, S.R.; Ma, R.; Ruan, L.; Todoroki, T.; Tang, Z.; Xu, Z.; Yang, C.; Yang, Q.; Yang, S. Coherent J/ψ photoproduction in hadronic heavy-ion collisions. *Phys. Rev. C* **2018**, *97*, 044910. [CrossRef]
97. Zha, W.; Ruan, L.; Tang, Z.; Xu, Z.; Yang, S. Coherent photo-produced J/ψ and dielectron yields in isobaric collisions. *Phys. Lett. B* **2019**, *789*, 238–242. [CrossRef]
98. Wang, P.; Wu, X.; Zha, W.; Tang, Z. Calculations of differential momentum transfer spectra for J/ψ photoproduction in heavy-ion collisions. *Chin. Phys. C* **2022**, *46*, 074103. [CrossRef]

99. Kłusek-Gawenda, M.; Szczurek, A. Photoproduction of J/ψ mesons in peripheral and semicentral heavy ion collisions. *Phys. Rev. C* **2016**, *93*, 044912. [CrossRef]
100. Zha, W.; Ruan, L.; Tang, Z.; Xu, Z.; Yang, S. Coherent lepton pair production in hadronic heavy ion collisions. *Phys. Lett. B* **2018**, *781*, 182–186. [CrossRef]
101. Li, C.; Zhou, J.; Zhou, Y.J. Probing the linear polarization of photons in ultraperipheral heavy ion collisions. *Phys. Lett. B* **2019**, *795*, 576–580. [CrossRef]
102. Li, C.; Zhou, J.; Zhou, Y.J. Impact parameter dependence of the azimuthal asymmetry in lepton pair production in heavy ion collisions. *Phys. Rev. D* **2020**, *101*, 034015. [CrossRef]
103. Heisenberg, W.; Euler, H. Consequences of Dirac's theory of positrons. *Z. Phys.* **1936**, *98*, 714–732. . [CrossRef]
104. Harland-Lang, L.A.; Khoze, V.A.; Ryskin, M.G. Exclusive LHC physics with heavy ions: SuperChic 3. *Eur. Phys. J. C* **2019**, *79*, 39. [CrossRef] [PubMed]
105. ALICE Collaboration. Letter of intent for ALICE 3: A next-generation heavy-ion experiment at the LHC. *arXiv* **2022**, arXiv:2211.02491.
106. SN0755: The STAR Beam Use Request for Run-21, Run-22 and Data Taking in 2023-25 | The STAR Experiment. 2021. Available online: https://drupal.star.bnl.gov/STAR/starnotes/public/sn0755 (accessed on 1 September 2020).
107. Prospects for Measurements of Photon-Induced Processes in Ultra-Peripheral Collisions of Heavy Ions with the ATLAS Detector in the LHC Runs 3 and 4-CERN Document Server. 2018. Available online: http://cds.cern.ch/record/2641655/files/ (accessed on 4 October 2018).
108. ATLAS Collaboration. Measurement of muon pairs produced via $\gamma\gamma$ scattering in non-ultraperipheral Pb+Pb collisions at $\sqrt{s_{\mathrm{NN}}} = 5.02$ TeV with the ATLAS detector. *arXiv* **2022**, arXiv:2206.12594. [CrossRef]

MDPI AG
Grosspeteranlage 5
4052 Basel
Switzerland
Tel.: +41 61 683 77 34

Symmetry Editorial Office
E-mail: symmetry@mdpi.com
www.mdpi.com/journal/symmetry

www.ingramcontent.com/pod-product-compliance
Lightning Source LLC
Chambersburg PA
CBHW062051220926
51850CB00010B/2727
9783725842155